National Conference on the Role of Polish Coal in the National and European Energy Sector 2017

AA006802

IOP Conference Series: Materials Science and Engineering
Volume 268

Andrychow, Poland
19 - 21 June 2017

Editors:

Piotr Banka
Marcin Lutynski
Marta Matyjaszek

ISBN: 978-1-5108-5267-9
ISSN: 1757-8981

Printed from e-media with permission by:

Curran Associates, Inc.
57 Morehouse Lane
Red Hook, NY 12571

Some format issues inherent in the e-media version may also appear in this print version.

Printed by Curran Associates, Inc. (2017)

For permission requests, please contact the Institute of Physics
at the address below.

Institute of Physics
Dirac House, Temple Back
Bristol BS1 6BE UK

Phone: 44 1 17 929 7481
Fax: 44 1 17 920 0979

techtracking@iop.org

Additional copies of this publication are available from:

Curran Associates, Inc.
57 Morehouse Lane
Red Hook, NY 12571 USA
Phone: 845-758-0400
Fax: 845-758-2633
Email: curran@proceedings.com
Web: www.proceedings.com

Table of contents

Volume 268

The Role of Polish Coal in the National and European Energy Sector
19–21 June 2017, Andrychów, Poland

Accepted papers received: 14 November 2017
Published online: 28 November 2017

Preface

Papers

IOP Conf. Series: Materials Science and Engineering **268** (2017) 011001 doi:10.1088/1757-899X/268/1/011001

1. Preface

Dear Ladies and Gentlemen,

Between 19th and 21st of June 2017 in Andrychów (Poland) took place the national conference entitled 'Role of Polish coal in the national and European energy sector'. The conference gathered over 120 participants and among them major figures from the mining and energy sector in Poland both from academia and industry. Poland with its largest hard coal mining reserves in European Union struggles with carbon emission limits which in the near future will cause a significant problem for the energy sector and as a consequence – to the mining industry. This meeting was a place of exchange of ideas and views on the future of energy and coal industry in Poland. The following subjects were within the scope of this meeting:

- Brown and hard coal sector market analysis.
- Energy market analysis.
- Mining and energy sector development scenarios for Poland.
- Mining and energy production costs.
- Health and safety in mining industry.
- Modern methods of power generation from coal.
- Coal in the heating sector.
- Energy storage.
- Carbon capture and storage.

The conference series is a result of selection of the most important articles. All articles were peer reviewed through process administrated by the Editors.

Piotr Bańka Marcin Lutyński Marta Matyjaszek

IOP Conf. Series: Materials Science and Engineering **268** (2017) 011002 doi:10.1088/1757-899X/268/1/011002

Peer review statement

All papers published in this volume of *IOP Conference Series: Materials Science and Engineering* have been peer reviewed through processes administered by the proceedings Editors. Reviews were conducted by expert referees to the professional and scientific standards expected of a proceedings journal published by IOP Publishing.

IOP Conf. Series: Materials Science and Engineering **268** (2017) 012001 doi:10.1088/1757-899X/268/1/012001

Importance of hard coal in electricity generation in Poland

Franciszek Plewa and Grzegorz Strozik

Faculty of Mining and Geology, Silesian University of Technology, Akademicka 2, 44-100 Gliwice, Poland

grzegorz.strozik@polsl.pl

Abstract. Polish energy sector is facing a number of challenges, in particular as regards the reconstruction of production potential, diversification of energy sources, environmental issues, adequate fuels supplies and other. Mandatory implementation of Europe 2020 strategy in terms of "3x20" targets (20% reduction of greenhouse gases, 20% of energy from renewable sources, and 20% increase of efficiency in energy production) requires fast decision, which have to be coordinated with energetic safety issues, increasing demands for electric energy, and other factors. In Poland almost 80% of power is installed in coal fired power plants and energy from hard coals is relatively less expensive than from other sources, especially renewable. The most of renewable energy sources power plants are unable to generate power in amounts which can be competitive with coal fires power stations and are highly expensive, what leads o high prices of electric energy. Alternatively, new generation of coal fired coal power plants is able to significantly increase efficiency, reduce carbon dioxide emission, and generate less expensive electric power in amounts adequate to the demands of a country.

1. Introduction

First impetus to modernise energy production on global scale gave the Kyoto Protocol (1997), the treaty which has extended United Nations Framework Convention on Climate Change from 1992 and committed state parties to reduce greenhouse gases emissions. European answer for Kyoto Protocol has been European Strategy and Energy Union, which has been formulated at Climatic Conference in Paris 2015, and implemented in Europe 2020 Strategy. The headline targets of EU is reduction greenhouse gas emissions by at least 20% compared to 1990 levels, increase the share of renewable energy in final energy consumption to 20%, and achieve a 20% increase in energy efficiency [1, 2]. These ambitious objectives determine also targets of Polish energetic policy, however not to forgotten is that Polish energy industry is strongly dependent on fossil fuels and discussion about future of Polish energy market must also involve coal mining industry. The exploitation of coal mine deposits in underground mines is becoming increasingly difficult and costly. The fixed and variable costs borne by the coal mines are mostly dependent on the amortization costs of machines and devices, the costs of materials and energy, remunerations, external services, taxes and fees as well as other prime costs [3, 4].

- Costs of electric power include above all costs of energy raw material or – in more general terms – energy sources and costs of its production. A lot of components contribute to costs of electricity production, from which maybe distinguished [5]:
- Investment costs, which means capital required for building power plants, and return of capital,
- Fixed costs of power plants maintaining, which include costs of maintenance of the assets, renovations, fees and taxes, remuneration of Staff and other,

- Variable costs consisting of purchase costs of fuel, its delivery, and emissions allowances costs – in case of fossil fuels fired power plants.

From the point of view of energy recipients, price of energy depends also on costs of electric power grid maintenance, distribution, and energy losses in power energy networks.

Structure of primary energy sources used for production of electric energy (secondary energy) depends on economic indicators (criterion of the lowest prices), national security issues, environmental background, as well factors such as widely understood energy policy (energy development strategy) or use of differentiated primary energy sources for production of electricity.

Within increasing production of electric energy and differentiation of primary energy sources, which in large part can be seen as proportion between fossil fuels and renewable energy sources (RES), assessment of the role of hard coal in the structural transformations in electric energy production is of great importance.

2. Structure of primary energy sources and electricity demand in Poland

Accordingly to data from the end of 2014, slightly more than a half of Polish power engineering is coming from hard coal fired power stations, see table 1.

The projection suggests the electricity demand, at least in the range of dozen or so upcoming years, will be increasing. Upwards trend in consumption of electric energy in Poland will not differ from global economy perspectives.

Increasing demand for electric energy must be accompanied width adequate increase of its supply, what means that new investments in power industry sector are necessary. One can take under consideration that a lot of older power generation units must be gradually over time replaced by new ones due to their obsolescence.

To solve these problems investments in power generation branch will be required both for increase of electricity production and sustaining its current level. Forecasted demands for electric energy and restoration investments presents table 2.

Table 1. Power installed in Polish power plants (state on 31.12.2014) [5, 6].

Type of power plant	Electric power installed (GW)
Total	39 353
Professional heat power stations	31 087
- from which:	
on lignite	9 221
on hard coal	20 291
on natural gas	927
Professional hydropower plants	2 208
- from which pump storage	1 330
Industrial heat and power plants	1 871
Independent RES power plants	4 187
- from which wind power plants	3 866

Table 2. Forecasted increase in consumption of electric energy and investments in power industry [7].

Energy consumption/power capacity	Year			
	2015	2020	2025	2030
Predicted gross consumption of electric energy (TWh)	166.4	177.9	190.3	203.5
Total power capacity (installed) (MW)	38 000	39 500	41 700	44 700
Predicted total power capacity from power plants already existing in 2011 (MW)	34 500	28 700	24 700	17 800
New power capacity required (MW)	3 500	10 800	17 000	26 900

IOP Conf. Series: Materials Science and Engineering **268** (2017) 012001 doi:10.1088/1757-899X/268/1/012001

3. Costs of new power generation installation

Development of energy sector could follow different scenarios, accordingly to preferences toward different types of fuels and energy sources as well as factors influencing energy policy. Among them reduction of carbon dioxide to atmosphere will be crucial.

In Poland, where electricity production is based on hard coal and lignite, scenarios of power industry development may include only partial departure from coal as the main energy source. Hence, for all alternative concepts coal as the dominating energy source remains the reference point.

For the purpose of investing in new power plants, from the point of view of energy cost production, such ones should be considered, which will be incurred to obtain required power capacity in the plants. Graph on figure 1 shows estimated cost of acquisition of a unit of electric power for different energy sources.

From the data presented in figure 1 on may conclude that in terms of maximum (installed) power, the lowest expenditures require natural gas power plants, where installation of 1 MW power requires to incur expenditures in the amount of 3.9 million PLN. Second position with investment cost of 6.6 million PLN per megawatt is taken by coal fired and wind power plants. Only by one million PLN greater is the unit cost (PLN/MW) for photovoltaic plants. The highest cost must be incurred for investments in hydropower. It should be pointed out that in rough approximation costs of construction of the off-shore wind facilities are close to nuclear plants and twice higher as on-shore wind power plants and hard coal power stations.

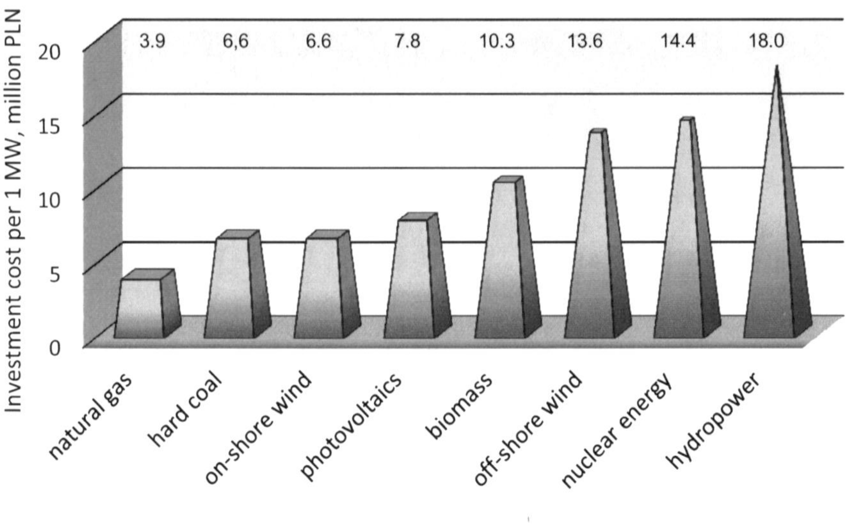

Figure 1. Construction costs of a power plant calculated for maximum power based on 2013 predictions [5, 8].

To asses real cost production of electric energy from different sources, utilisation rate resulting from limited period of time during a year, within power plant produces energy, must be considered. Average power as a statistical parameter is representing a ratio between installed power and number of days during year of power generation This parameter allows to assess actual cost of investment per unit of power for different energy sources.

Duration of work of generators depends on necessary maintenance breaks (e.g. renovations, exchange of fuel in nuclear plants) and – as it is the case for the most of RES – interruptions resulting from variability of meteorological conditions [5, 6]. The longest work time during a year characterizes power plants based on fossil fuels and nuclear energy, which are reduced only by periodical maintenance and replacement of nuclear fuel respectively [9].

Although large investments in development of nuclear energetics would provide relatively inexpensive electric energy in large amounts, this direction of power industry development seems to be unrealistic both due to existing policy of power generation development and negative social perception of nuclear plants construction. Building of nuclear power plants still remains in doubt and in predictable future it would be hardly to expect a point of turning, especially considering a global trend to get out of this kind of energy source. Advantageous feature of nuclear power implementation is long life time of a plants, which is about 60 years long in contrary to 40 years of exploitation of an coal fired plant [9].

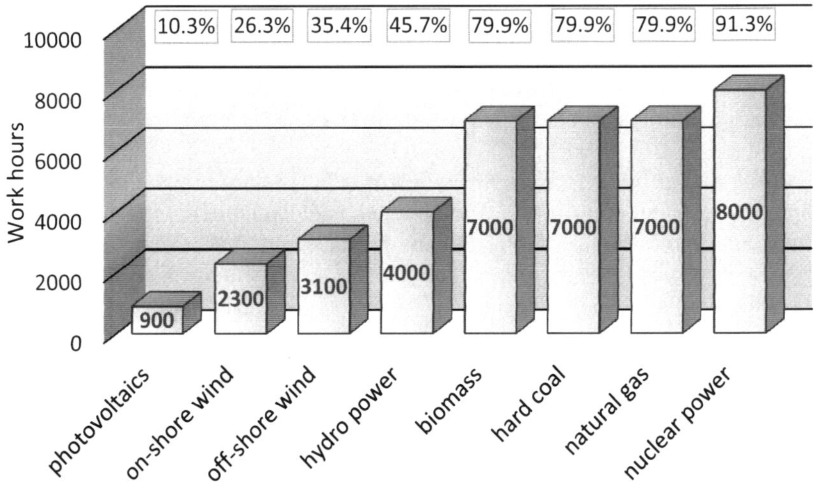

Figure 2. Time of use of the power installed in different types of power stations [5, 6].

Weather conditions in Poland (variability of meteorological conditions, relative small number of sunny and windy days) are disadvantageous enough, to use full capacity of installed power only in about 10% of a time yearly for solar energy and 23% - 35% for wind based generators. Absence of guaranteed deliveries of energy from providers operating in area of RES forces national energy system to keep running certain amount of power installed in coal fired power stations, for the purpose of compensation of energy losses generated by RES based facilities. This is especially important while the most available RES in Poland is energy of wind [10].

The lowest investment costs to produce 1 MW of power gain natural gas power plants (3.9 million PLN/MW), then hard coal fired power plants and on-shore wind plants with the cost of 6.6 million PLN/MW. Only little higher expenditures are required by photovoltaics, where installation of 1 MW costs 7.8 million PLN/MW. Usage of other energy sources is much more capital-intensive.

From the point of view of costs of electric energy, low investment cost of power station construction is fading its significance in the face of high natural gas prices and exposing the country to a risk of dependence on import of this fuel.

Comparing data shown in figures 1 and 3 one can see that the lowest investment cost, in relation to obtained average power, still relates to natural gas (4.9 million PLN/MW). However, one should keep in mind that increasing consumption of natural gas must be in 100% covered by its importing.

The second lowest cost of acquisition of average power represents hard coal (8.3 million PLN/MW), on the other hand the most expensive average power being is the one being generated from RES. Within this group of resources especially expensive venture is instalment of photovoltaic cells, which are able to generate power only by 900 hours yearly, thereby their average power is 11 times smaller than installed [5].

Also inviable seem to be investments in hydropower, which represents the second highest cost of average power acquisition. It should be also mentioned, that water dams are in use mainly for the purposes of flood protection and power plants accompanying them are of less importance.

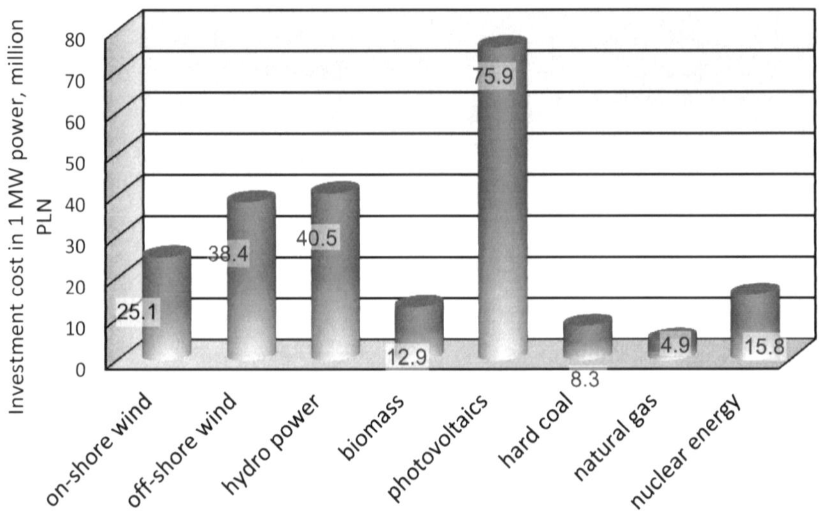

Figure 3. Investment costs in RES calculated as million PLN/MW of average power accordingly 2013 prices.

On the basis of investments values and operational costs (including purchase of fuels), rough costs of energy unit may be assessed. Table 3 presents production costs of 1 MWh of electric energy in Poland for selected methods of its generation.

Definitely most expensive energy source is undoubtedly photovoltaics with production cost of 1091 PLN/MWh. High investment costs are accompanied by high operational costs (maintenance), which amount about 10% - 15% of investment value for small installations of less than 1.5 kW power, down to 1% - 2% for installations between 100 kW and 1 MW [11].

Table 3. Production costs of electric energy in Poland from different sources [8, 6, 12].

Energy source	Production cost (PLN/MWh)	Energy source	Production cost (PLN/MWh)
On-shore wind	466	Photovoltaics	1091
Off-shore wind	713	Hard coal	282
Hydropower	484	Natural gas	314
Biomass	784	Lignite	171
Nuclear power	313		

Nowadays as well as in near future, electric power from hard coal combustion still is and will be the least expensive one. Even by increasing cost of its production resulted from emission allowance fees, which will make it closer to other energy sources, it will remain the most cost-effective [13].
Currently the percentage of different kinds of RES in production of electric energy in Poland amounts respectively: 10.7% hydropower 43.1% wind, 41.0% biomass, 0.2% photovoltaics, and 5% other sources [4]. In previous years wind turbines became much less expensive (by 25% in period 2008 – 2014), but it cannot be expected further decreasing of their prices, which could reduce costs of their instalment and energy [5].

By such comparisons one should keep in mind that each energy source is characterized by different scale of power, which it is able to provide to the total energy balance at the level of demands of the whole country. In this approach all kinds of RES may be consider as useful on a local level, where no gross consumer occur. Dispersion of power plants powered by RES (mainly small wind turbines) is advantageous in this context on rural areas, where delivery of electric energy from large power stations is connected with high energy losses by its supply. Average electric losses in supply reach up to 9% of energy generated.

Wind power plants, frequently in a form of single turbines, are built in areas, which ensure possibly highest number of windy days. Additionally, differentiation of meteorological conditions over the country should allow continuous delivery of energy from at least a certain part of them, however even these factors do not guarantee stable level of power coming from wind. For instance in July of 2015 a critical situation occurred in national power balance, when due to still air weather over the whole country, wind power stations generated only 430 MW from 4178 MW of power standing potentially for disposition. Considering recent legislation changes it is hardly to count on further development of on-shore wind power stations. However, an idea of construction the first off-shore wind farm has appeared, which is targeted to possess 600 MW of installed power by expenditures of about 10 billion PLN that means almost 17 million PLN per megawatt of power installed [10].

4. Production costs of electric energy in other countries

Production costs of electric energy from hard coal in Poland, Germany, and U.S. are similar and amount 282 PLN/MWh, 280 PLN/MWh, and 300 PLN/MWh respectively. Cost of production of energy from on-shore wind generators in Poland with 466 PLN/MWh is significantly higher than in Germany and U.S., where these costs are similar and amount 280 PLN/MWh and 260 PLN/MWh respectively [14, 15, 16].

On the example of data from United States, the influence of technology applied in hard coal fired power plants on energy production costs can be demonstrated. Accordingly data collected in Table 4, the least expensive energy comes from power plants equipped with steam-gas turbines (round 66 – 67 USD/kWh), though cost of carbon dioxide capture and storage are estimated at the level of 12.5 – 27.8 USD/kWh. Accordingly presented data, energy from nuclear power plants may be less expensive than from hard coal fired power stations, what is also probable in Poland, while the production cost of 1 MWh energy from hard coal provided by table 3 does not consider differentiation of power generation technologies.

Table 4. Estimated production costs of energy in coal fired power plants commissioned after 2018 in U.S. (USD/MWh) [15].

Type of technology	Time of work (%)	Return of investment	Fixed operational costs	Variable operational costs	Energy transfer costs	Energy production cost
Coal – ordinary	85	65,7	4.1	29.2	1.2	100.1
Coal – advanced	85	84.4	6.8	30.7	1.2	123.0
Coal – Advanced + CCS*)	85	88.4	8.8	37.2	1.2	135.5
Gas – ordinary SGT**)	87.8	15.8	1.7	48.4	1.2	67.1
Gas – advanced SGT	87	17.4	2.0	45.0	1.2	65.6
Gas – advanced SGT and CCS	87	34.0	4.1	54.1	1.2	93.4
Nuclear	90	83.4	11.6	12.3	1.1	108.4

*) CO_2 capture and storage, **) Steam-gas turbine

In terms of RES in American conditions, the least expensive and additionally the most competitive in regard to hard coal fired power plants are on-shore wind power plants, what can arguable resulted from climatic conditions [15].

On the background of data presented in table 5, the most expensive sources of electric energy are off-shore wind power plants and photovoltaics (similarly as in Poland). Especially expensive is production of electricity from solar thermal energy, which is still at the level of a few pilot plants worldwide.

In turn, production costs of energy in Germany have been collected in figure 4. They stay in general proportion do energy production costs in Poland, except photovoltaics, for which electric energy production cost is comparable with natural gas and on-shore wind plants.

Table 5. Estimated production costs of electric energy from RES after 2018 in U.S. (USD/MWh) [15].

Type of RES	Time of work (%)	Return of investment	Fixed operational costs	Variable operational costs	Energy transfer costs	Energy production cost
Wind – on-shore	34	70.3	13.1	0.0	3.2	86.6
Wind – off-shore	37	193.4	22.4	0.0	5.7	221.5
Photovoltaic	25	130.4	9.9	0.0	4.0	144.3
Thermal solar	20	214.2	41.4	0.0	5.9	261.5
Hydropower	52	78.1	4.1	0.0	2.0	90.3
Biomass	83	53.2	14.3	6.1	1.2	111.0
Geothermics	92	76.2	12.0	42.3	1.4	89.6

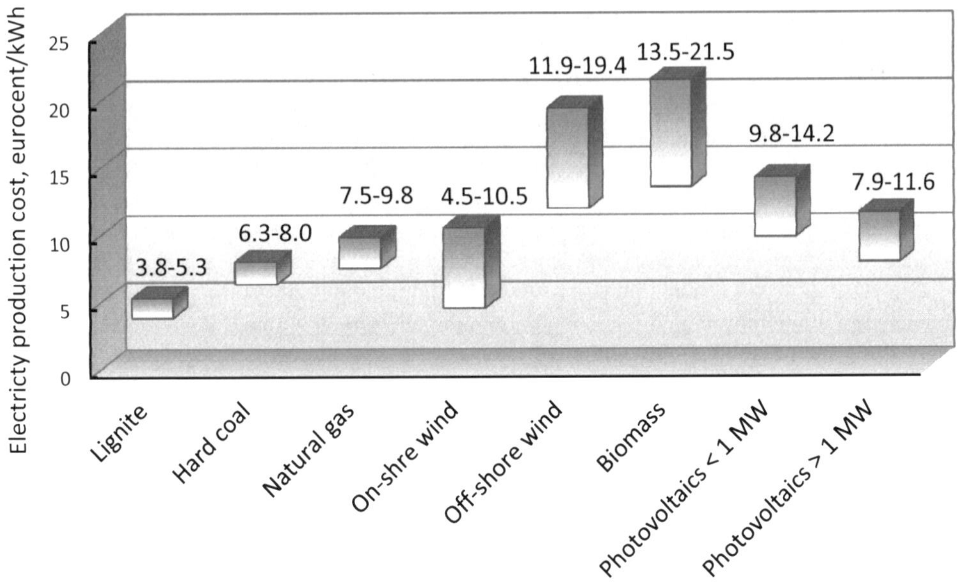

Figure 4. Electric Energy production costs in Germany in 2013 [16].

5. Global trends in share of hard coal in production of electric energy

Share of coal (both hard coal and lignite) in global production of electric energy is not dominating and amounts round 39.5% [7]. However for many world leading economies it is primary or having high share source of energy for production of electricity. Share of coal in production of electric energy in selected countries illustrates the chart in figure 5. Coal remains an important source of energy production in all countries, which are its largest producers. However, most of these countries decreased production of electric energy. Between 1995 and 2015 Poland's share of coal as a source of electricity decreased only from 97.5% to 80.9%, while in U.K. from 81.1 to 22.9%, or in USA. from 53.9% to 34.3% [19]

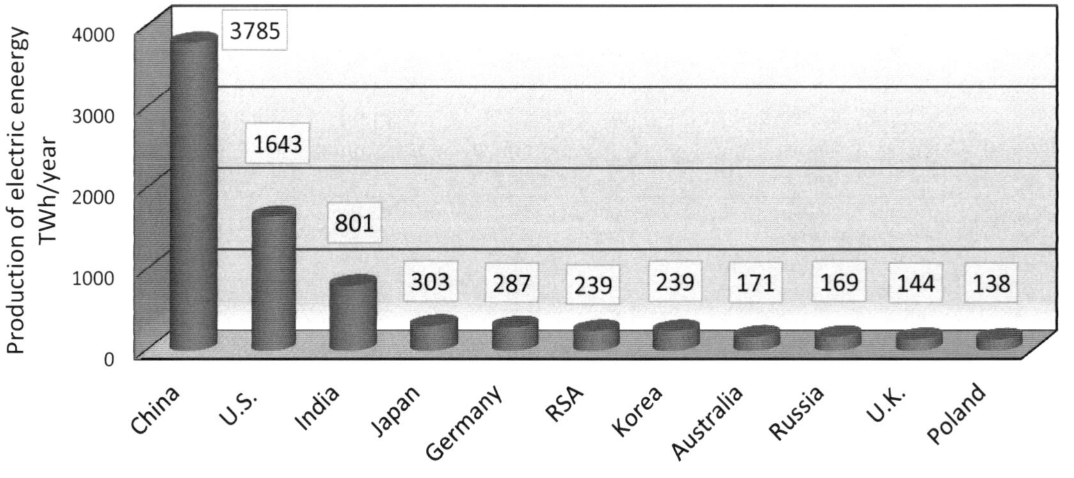

Figure 5. Shares of coal in energy production by selected countries [10].

Share of coal as well as other fuels and energy sources in production of electricity undergoes long-term trends resulting from changes of prices (e.g. oil), technical development (e.g. nuclear power and RES), or environmental policies (e.g. coal and RES). In the whole period taken under analysis in figure 5, decreasing trend in use of oil can be observed since decade of 70's of 20th Century, leading to marginal share of oil consumption for electric energy production. Use of natural gas since last years is almost constant and should not change significantly until 2035 [17]. Also nuclear power after a period of increased interest, will keep constant share in production of electric energy. In turn, hydropower demonstrates constant share over the whole considered period, what could suggest lack of development opportunities, due to high investment costs and objections from the communities, which resist building of hydropower stations, due to huge social and environmental implications of creating large artificial water reservoirs.

Accordingly the forecast of shares structure of electric energy sources shown in figure 6, RES are representing relatively new source of energy, which development within time span up to 2035 is limited and may achieve about 12% of share in total energy production in relation to current share of only 5%.

Consumption of hard coal in global scale has been significantly increased since 2000 and this trend will last up to about 2020 (figure 7). Increased demand for hard coal is an implication of sudden and significant economic progress in China. Also economy of India indicates strongly growing demand for hard coal. In case of Europe demand for hard coal will decrease by 50% in perspective until 2035 [18]. However the forecast shows clearly that in global scale demand for hard coal will be increasing, mainly due to countries, which are expected to step into progress phase slightly later.

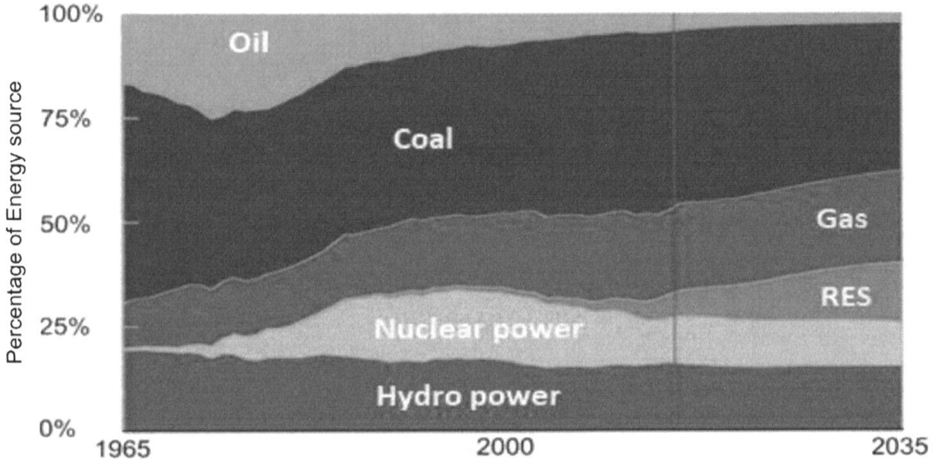

Figure. 6. Global shares of electric energy sources [19].

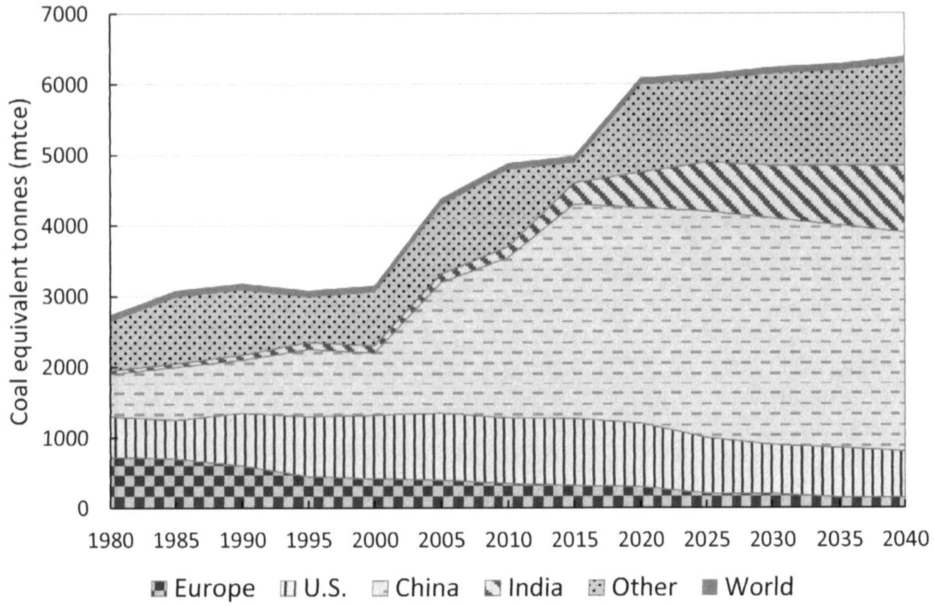

Figure 7. Forecast of demands for hard coal by world regions [19].
[coal equivalent: mass of coal, which combustion generates 29.39 GJ (8.14MWh) of energy]

6. Production of electric Energy in Poland at the background of EU

A suitable example illustrating possible changes in energetic economy in Poland might be Germany, where similarly as in Poland production of electric energy is based on hard coal and lignite. Simultaneously Germany create broad field for observation of RES problematics.

In 2016 Germany produced total amount of 648 million kWh of electric energy. From that 338 million kWh (52%) came from fossil fuels, and 30% (191 million kWh) from RES (figure 8). Source of 40% of produced electric energy was hard coal and lignite. Between RES the most important energy source is wind with 12.3% share in total production of electric energy [3, 4].

In the same year in Poland has been produced 164 million kWh of electric energy, from which 22 million kWh came from RES, making 13,43% of total electric energy produced [19]. Thereby, share of RES in Polish electricity production doubled since 2010.

Prices of electric energy in Poland belong to medium between EU member countries and in 2016 amounted 3.52 cent/kWh for households and 7.1 cent/kWh for gross consumers [22]. The most expensive is the electric energy in Denmark and Germany, where its price for households amounted 30.84 cent/kWh and 29.77 cent/kWh respectively [20]. At the same time in these countries share of RES in production of electric energy is the highest among EU member countries with 30.0% and 51.3% respectively, what seems to be a non-accidental correlation, particularly in view of the fact that total amounts of electricity production in those countries are incomparable, while Denmark produces 17 times less electricity than Germany [3].

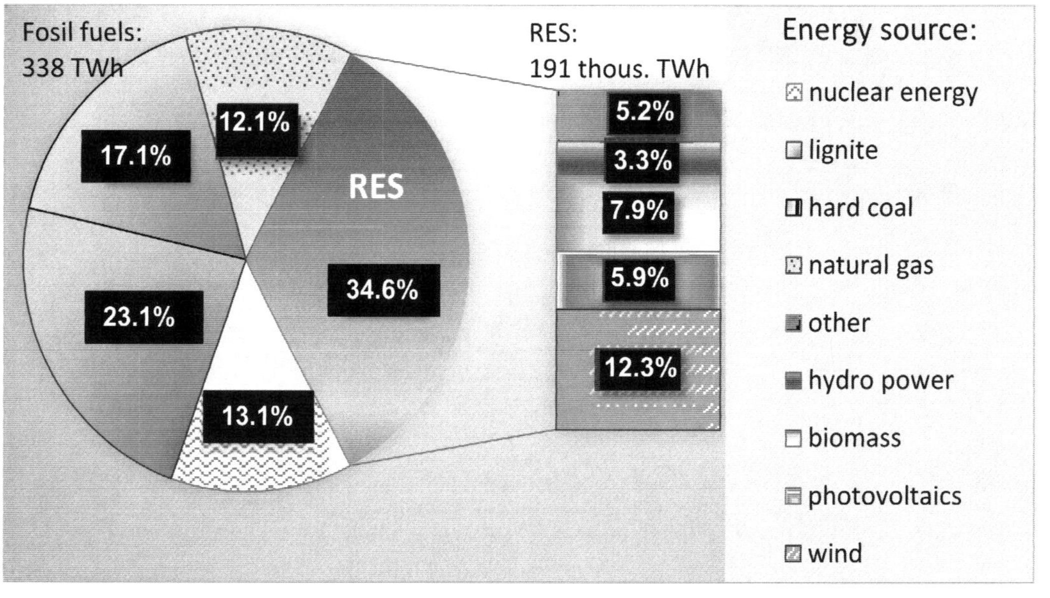

Figure 8. Share of particular electric energy sources in Germany at the end of 2016 [3].

Pursuant to the EU guidelines, until 2020, 20% of electric energy being produced in Poland should come from RES. Thereby share of RES must increase from 13.43% in 2015 up to 20% in 2020 [4] at the expense of non-renewable sources, mainly hard coal and lignite, which share should be limited by about 6% [21].

However taking under consideration forecasted general increase of electric energy demand, estimated as 7% between 2015 and 2020, demand for energetic coal might stay at the same level. In such a way influence of increasing percentage of electricity from RES on the demand for hard coal and lignite in longer perspective should be at least partially compensated.

7. Emission of carbon dioxide in coal based power generation

Figure 9 shows carbon dioxide emissions to the atmosphere in Poland and Germany in years 2005 - 2014. Before 2010 mass of CO_2 emitted to the atmosphere was decreasing and during further years followed constant level of round 820 million tonnes in Germany and 325 million tonnes in Poland per year [7].

Because of the size of the economies in Poland and Germany this difference is not surprizing, although interesting could be calculation of CO_2 emission per capita (table 6). In such an approach, emission of carbon dioxide amounted in considered years round 10.3 tonnes in Germany and 8.4 tonnes in Poland per inhabitant.

Accordingly Eurostat [3] indicator of greenhouse gases emission calculated as energy being produced with emission of CO_2 to the total amount of produced energy on 2000 basis (100%), took in 2014 value of 95.8% in Germany and 90,1% in Poland by average for EU members countries 89.1%. From the point of view of these data, climate doctrine of German environment protection policy should be more concentrated on giving a positive example rather than pressure on Poland and other EU member countries to reduce CO_2 emissions.

Implementation of the carbon dioxide emissions charges and building of CO_2 capture and storage plants will be reflected in higher production cost of electric energy. For instance, in case of lignite fired power plant, implementation of CO_2 capture and storage facility will increase production cost of 1 MWh energy up to round 280 ÷ 300 PLN.

In turn, by predicted increase of carbon dioxide emission cost up to 25 ÷ 30 EUR/tonne, production cost of electric energy from lignite would reach a level of about 270 ÷ 290 PLN/MWh and 380 ÷ 400 PLN/MWh for hard coal. Besides higher energy prices, increased production costs would give an impulse to reduce CO_2 emission as result of its capture and storage and creation of better environment for development of RES based energetics [8].

Table 6. Carbon dioxide emissions in Poland (PL) and Germany (D) from 2010 to 2014 [7]

CO_2 emission	2010		2011		2012		2013		2014	
	PL	D	PL	D	PL	D	PL	D	PL	D
million t/year	828.2	324.5	802.3	337.0	815.0	326.0	842.0	321.1	798.6	316.8
t/year per capita	9.32	7.99	9.9	9.1	10.0	8.5	12.7	8.3	9.8	8.2

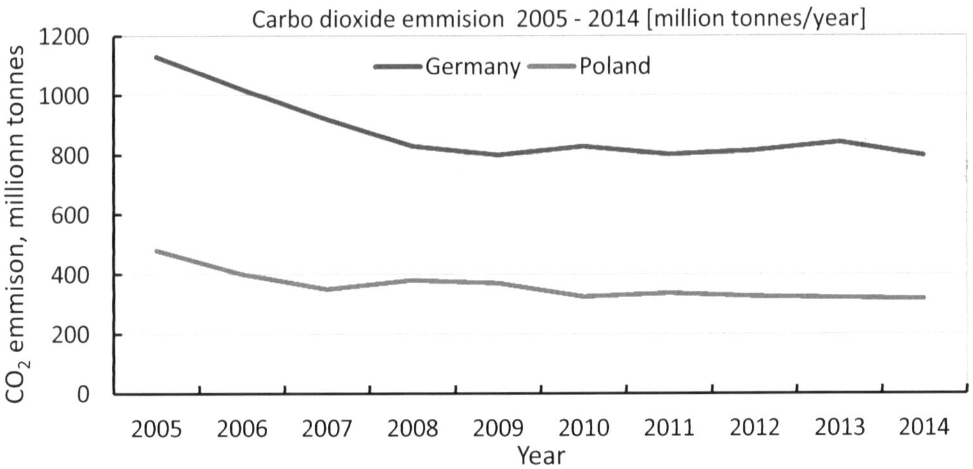

Figure 9. Carbon dioxide emission in Poland Germany in period 2010 – 2014 [7].

8. New coal fired power plants in Poland

In the framework of restoration and energy potential development investments in power industry new power plants are built as well as old power units are replaced with new ones. These new installations ensure much higher efficiency and smaller carbon dioxide emission, which is reflected in lower coal consumption and smaller CO_2 emission per unit of energy [22].

Hard coal fired power stations with conventional energy units are working on sub-critical parameters with performance of about 37%, however efficiency of older, exploited units is much smaller. For instance, efficiency of power units 1 – 4 in power plant Rybnik is about 34% ÷ 35%, while the

IOP Conf. Series: Materials Science and Engineering **268** (2017) 012001 doi:10.1088/1757-899X/268/1/012001

performance of new units for sub-critical parameters reaches 38% ÷ 42% and in the latest units operating on super-critical parameters reaches 44 ÷ 46% [13]. These units will also have significantly reduced emissions of CO_2. The largest investments in new power units have been characterized in table 7.

Projects, which have been described in table 7 present only a selection from a plan of maintaining and development of electric energy production in Poland presented graphically in Figure 10. Power units and stations presented on the map contain objects of differentiated importance for national balance of energy production and probability of their implementation (at least in near future). Nevertheless one can expect that the majority of electric energy will be generated by power units of high output, fired with hard coal.

Table 7. Selected new investments in power units in Poland [7]

Power plant	Operator	Power, efficiency, start of building	Fuel
Power plant Turów (unit 11)	PGE	Power: 430 - 450 MW Efficiency 43.4% Start of building 06.2016.	Lignite
Power plant Opole (units 5 and 6)	PGE	Power: 2 x 900 MW Efficiency 45.5% Start: 2017/18	Hard coal
Power plant Jaworzno	Tauron	Power: 910 MW Efficiency: 45.9% Start: 0.6.2019.	Hard coal
Power plant Kozienice	Enea	Power: 900 - 1000 MW Efficiency 45.6% Start: 12.2017.	Hard coal

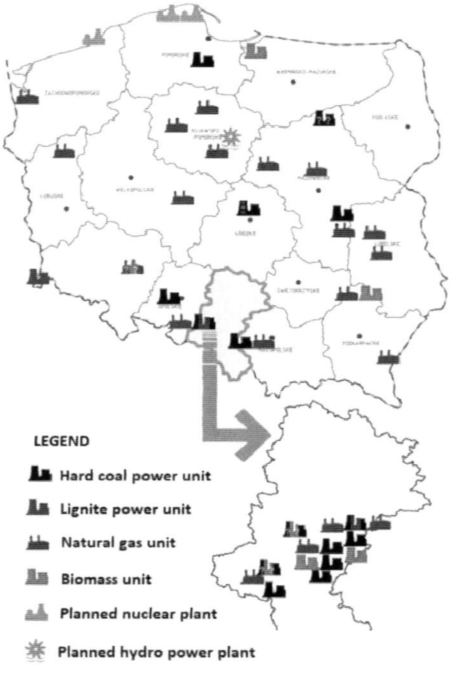

Figure 10. Investments in new power units and power stations in Poland [7].

The picture of evolution of Polish electric energy branch may be compared with current state and forecasts for power generation industry in Germany.

Despite declared concentration on RES, German energy industry concentrates first of all on coal as a fuel for new power plants intended to replace old plants and cover increasing demand for electric energy.

In the period of years 2014 – 2108 almost 12 GW of power from non-renewable sources will be shut down. Slightly more than a half of this power will be replaced by new power unit and power plants operating on fossil fuels (figures 11 and 12). Analysing structures of energy sources in deactivated and starting power stations one can see no change in amount of power generated in lignite fired power plants, which produce almost a quarter of total electric energy in Germany.

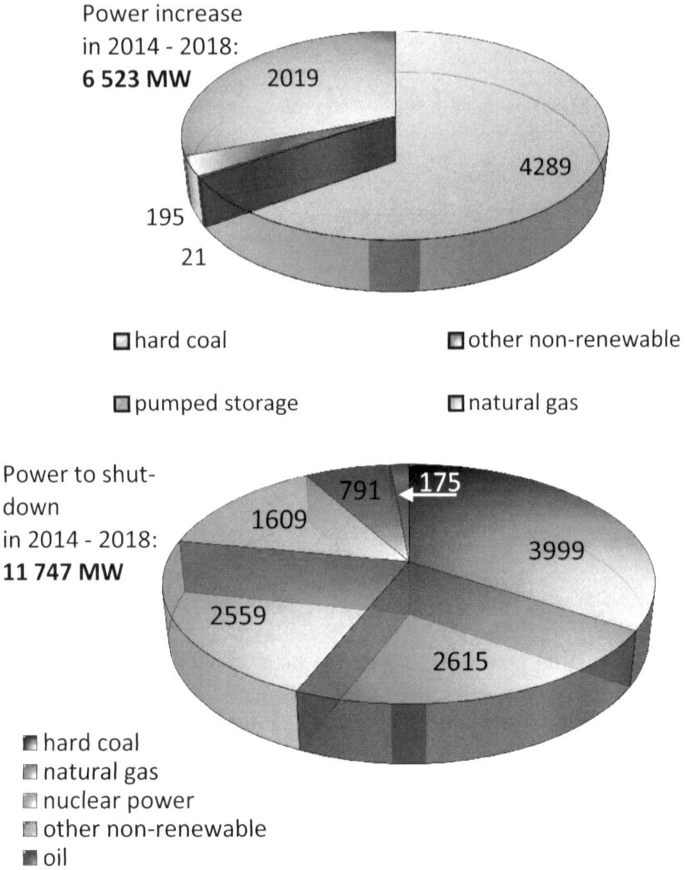

Figure 11. Plans of start-up and withdrawal of electric energy generation power in Germany [20].

In case of hard coal fired power plants, between 2014 and 2018, share of hard coal in production of energy will increase by 7%, taking finally share of 19% between all electric energy sources. Therefore, progress in RES based energy production had been used mainly in aim to replace production of energy from most expensive and import-dependent sources like oil and natural gas, as well as to make possible shut-down of negatively perceived nuclear power plants.

It should be emphasized that data from electricity production in Germany have been elaborated in years 2013 – 2015 and cover a perspective up to 2018, therefore the most of described tasks are already completed. In turn large number of coal fired power units designed to start in longer time perspective, what is illustrated in figure 11, indicates that reduction of coal consumption of electric power generation does look unrealistic

IOP Conf. Series: Materials Science and Engineering **268** (2017) 012001 doi:10.1088/1757-899X/268/1/012001

In first place on power generation efficiency with performance of 61,5% are gas-steam turbines. However they must be powered with coal and natural gas [21]. In future should be available such units powered only hard coal with use of coal gasification process for generation of gaseous fuel [22].

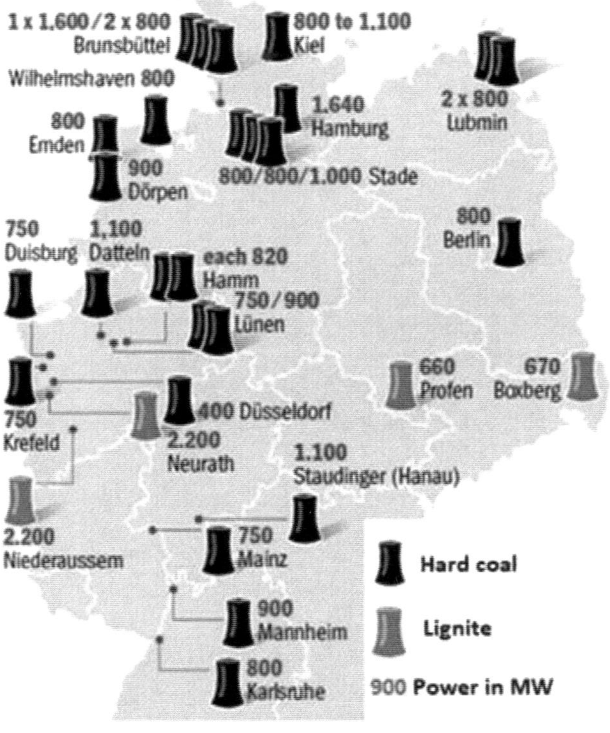

Figure 12. Planned new coal fired power plants in Germany [23].

9. Comments on development of coal production for inland electricity production in Poland

Although demands for energetic coal in perspective up to 20 years will not be a subject of significant decrease, economic condition of Polish hard coal industry is getting worse. In the most pessimistic scenario of changes in energy sources, is possible that hard coal would not be ousted by competition, but just replaced necessarily by other fuels and energy sources, including imported hard coal.

Restructuring of old mines in terms of ownership and accessing new not yet mined coal beds does not give guarantee the maintaining necessary level of coal production and particularly solid, long term competitiveness of prices.

On the areas of Uppers Silesia and Lublin coals basins exist favourable located, yet not mined reach resources. There is a need to build new mines on such areas, which activity from exploration, through building up to equipment and environment production should be achieved completely through Polish companies. A great asset in development of modern coal mining shall be undoubtedly high level of Polish mining technologies and know-how.

Not without significance for building of new coal mines are such factors as:

- Possibility to locate the mines in advantageous mining-geological conditions,
- Design of mine structure and access to the coal beds from the point of view of application of modern technical and organizational solutions,
- Best conditions to minimize impacts on the environment,

which influence heavily final price of energy fuel offered on the market.

IOP Conf. Series: Materials Science and Engineering **268** (2017) 012001 doi:10.1088/1757-899X/268/1/012001

10. Conclusion

Both global as well as European trends in source structure changes in power industry, despite gradual increase of RES participation, does not exhibit drastic decrease of demands for energetic hard coal. Growing demand for electric energy and necessity of old power units replacement, require significant investments in the power industry.

Examples of Poland and Germany shows that serious losses of power resulted from closure of old power stations may be compensated only by new coal fired power plants.

New power units are characterized by slightly higher efficiency and will be significantly more environmental friendly.

Production costs of electric energy will be influenced by buying of emission allowances and implementation of carbon dioxide capture and storage systems, however increased electric energy prices might be, at least in a part, compensated by achievement lower level of hard coal prices resulted from launching of new coal mines

References

[1] EUROPE 2020 - A strategy for smart, sustainable and inclusive growth (Annex 1 – Europe 2020: an Overview). EUR-Lex. European Commission. Retrieved 25 September 2015
[2] Raines T nad Tomlinson S 2016 Europe's Energy Union. Foreign Policy Implications for Energy Security, Climate, and Competitiveness, Europe Programme/Energy, Environment and Resources, Gatham House
[3] http://ec.europa.eu/eurostat/web/products-datasets/-/med_eg30
[4] http://ec.europa.eu/eurostat/web/energy/data/shares
[5] http://ncbj.edu.pl/niezalezna-ocena-kosztow-produkcji-energii-elektrycznej#podsumowanie
[6] Kwartalnik Agencji Rynku Energii *Sytuacja w energetyce* http://www.rynek-energi-elektrycznej.cire.pl
[7] http://euracoal.eu
[8] http://info.wyborcza.biz/szukaj/gospodarka/cena+za+tonę+emisji+co2
[9] Michalak J 2012 Analiza porównawcza opłacalności inwestycji węglowych i jądrowych *Polityka energetyczna* **15/4** pp 187-99
[10] https://data.worldbank.org/indicator/EG.ELC.COAL.ZS
[11] http://solaris18.blogspot.com/2012/09/koszty-montazu-i-eksploatacji.html
[12] Zaporowski B 2008 Analiza kosztów wytwarzania energii elektrycznej *Polityka energetyczna* **11/1** pp 531-42
[13] Zaporowski B 2011 Efektywność energetyczna i ekonomiczna elektrowni i elektrociepłowni dużej i średniej mocy *Polityka energetyczna* **14/2** pp 455-68
[14] https://de.wikipedia.org/wiki/Stromgestehungskoste
[15] Levelized Cost of New Generation Resources in the Annual Energy Outlook. U.S. Energy Information Administration, April 2017
[16] Stromgestehungskosten Erneubare Energien. Fraunhofer-Institut für Solare Energiesysteme ISE, November 2013
[17] Katzer J (ed.) 2007 The Future of Coal, Massachusetts Institute of Technology
[18] Finkelman R B and Shifenk D Coal as promising source of critical elements. Progress and future prospects *International Journal of Coal Geology Elsevier* Available online 19 June 2017
[19] BP Energy Outlook 2017 Edition
[20] Data on energy price trends. Long-Time Series from January 2000 to March 2017. Statistisches Bundesamt 2017
[21] Heberg A 2015 Europe'/s energy security – od the Energy Union the answer, European Policy Centre
[22] Ministerstwo Skarbu Państwa 2017 Przyszłość polskiego węgla jako surowca energetycznego
[23] Friends of the Earth: http://www.foeeurope.org/search/foee/new%20coal-fired%20power%20stations?page=1

IOP Conf. Series: Materials Science and Engineering **268** (2017) 012002 doi:10.1088/1757-899X/268/1/012002

Access to primary energy sources - the basis of national energy security

Jan Szlązak and Rafał A Szlązak

DG SCORPION, Fabryczna 15, 40-611 Katowice, Poland

dg.scorpion@dgscorpion.pl

Abstract. National energy security is of fundamental importance for economic development of a country. To ensure such safety energy raw material, also called primary energy sources, are necessary. Currently in Poland primary energy sources include mainly fossil fuels, such as hard coal, brown coal, natural gas and crude oil. Other sources, e.g. renewable energy sources account for c. 15% in the energy mix. Primary energy sources are used to produce mainly electricity, which is considered as the cleanest form of energy. Poland does not have, unfortunately, sufficient energy sources and is forced to import some of them, mainly natural gas and crude oil. The article presents an insightful analysis of energy raw material reserves possessed by Poland and their structure taking account of the requirements applicable in the European Union, in particular, those related to environmental protection. The article also describes demand for electricity now and in the perspective of 2030. Primary energy sources necessary for its production have also been given. The article also includes the possibilities for the use of renewable energy sources in Poland, however, climatic conditions there are not are not particularly favourable to it. All the issues addressed in the article are summed up and ended with conclusions.

1. Introduction

Energy security of countries has always been a priority for each government. Since the beginning of armed conflicts in North Africa and the Middle East, in countries with rich deposits of crude oil and natural gas - the issue of energy security of many countries, including Poland, makes the front pages of newspapers.

Energy security is not clearly defined. In the in the Energy Law Act of 10 April 1997 [1], the following definition, which is also used in this article, is given:

'National energy security – the condition of the economy which enables full coverage of the customer's ongoing and prospective demand for fuels and energy in a technically and economically justified manner, with the observance of the environment protection requirements'.

Demand for final energy and quantities of primary energy sources necessary to produce it are published in a government document entitled: 'The assumptions of the Polish energy policy until...'.

The current document sets out the perspective until 2030 and demand for energy raw materials even until 2050 (draft of policy assumptions ...) [2].

The production process of final energy, which is electricity, is schematically presented in figure 1. Electricity is considered as the cleanest energy and the most refined. Therefore, economic development of a country is determined on the basis of per capita energy consumption.

In the further parts of the article the said primary energy source will be discussed. Their usefulness for further development of the Polish economy will be assessed.

IOP Conf. Series: Materials Science and Engineering **268** (2017) 012002 doi:10.1088/1757-899X/268/1/012002

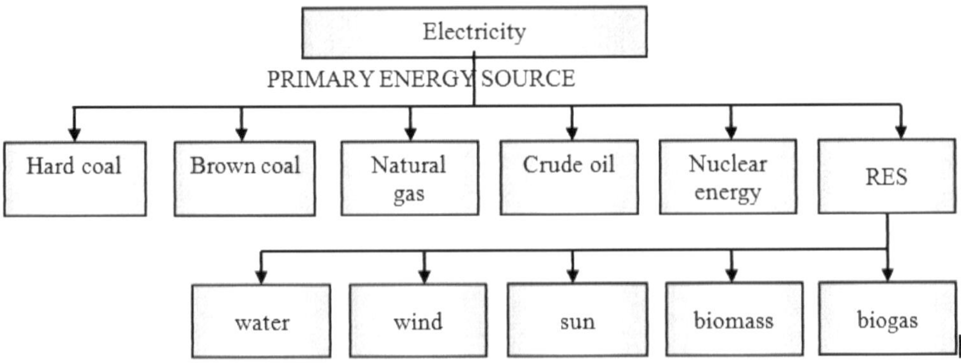

Figure 1. Process of electricity production.

2. Fossil sources of primary energy and their location in the territory of Poland

The distribution of hard coal and brown coal deposits in Poland is shown in figure 2. The following colours mean respectively:

- black - hard coal,
- brown - brown coal.

Źródło: Państwowy Instytut Geologiczny

Figure 2. Deposits of hard coal and brown coal in Poland.

Table 1 presents the balance reserves of the said energy raw materials, the number of the developed deposits and production as at 31 December 2015. The data presented in table 1 show that the balance reserves of hard coal increased by 4.26 billion Mg, despite its production of c. 72 million Mg in 2015. The increase of reserves results, however, mainly from the reassessment of coal reserves based on new criteria.

The reserves of other raw materials do not show significant changes, except for the changes resulting from production (gas, crude oil). It is worth noting that the gas balance includes also methane extracted from hard coal deposits, which until now has not been taken into account.

Table 1. Summary of geological balance reserves and production of important fossil fuels in Poland in 2015 - in million metric tons; natural gas and methane in billion metric tons; crude oil and gas - deposits extracted (Source: own study based on data of Polish Geological Institute).

Fossil fuel	Number of deposits			Balance reserves			Production	
	total	developed quantity	2014 =100%	situation on: 31/12/2015	including developed reserves	+ increase - decrease	volume	2014 =100%
ENERGY FOSSIL FUELS:								
- gas	352	236	101	213.59	138.09	-0.73	5.53	99.64
- liquid	86	64	96	22.82	22.26	-0.71	0.90	97.83
- solid	247	60	100	79 736.67	22.525.75	+4265.65	128.20	98.64
Natural gas	292	207	100	122.82	101.68	-4.70	5.21	99.05
Coal deposit methane	60	29	104	90.77	36.41	+3.97	0.32	110.34
Crude oil	86	64	96	22.82	22.26	-0.71	0.90	97.83
1	*2*	*3*	*4*	*5*	*6*	*7*	*8*	*9*
Brown coal	91	9	100	23.516.19	1418.70	+5.60	63.13	98.64
Hard coal	156	51	100	56.220.48	21.107.05	+4260.05	65.07	98.64

2.1. Hard coal

This raw material is the main source of primary energy used for electricity production. Over 50% of electricity has been produced using coal. Of all the energy sources at the disposal of the Polish energy industry, only the resources of hard coal and brown coal are sufficient. Therefore, the Polish energy industry is based mainly on those materials.

However, the reserves of those materials are determined in years. Hard coal reserves in Poland are shown in table 2 [4].

Table 2. Hard coal reserves in operating deposits as at 01/01/2015 (million Mg).

	Balance reserves	**Industrial reserves**	**Operable reserves**
During concession term	14.872	3.230	1.813
For entire deposit	20.282	6.023	3.390

Taking account of operable reserves of 3 3.390 million Mg and current production of c. 70 million Mg annually - the reserves are sufficient to meet c. 48 years of production, i.e. to c. 2070, if decreased production in the final stage of coal deposits mining is taken into account.

Hard coal reserves in Poland compared to selected countries are shown in table 3 [5].

Table 3. World hard coal production between 2008 – 2016 (in thousand Mg) Source: GUS (Central Statistical Office).

COUNTRY	2008	2009	2010	2011	2012	2013[a]	2014	2015	2016
World	5.688.416	5.789.687	5.877.025	6.235.943	6.454.799	7.240.791	7.212.081	.	.
Australia	431.004	426.996	449.004	443.700	488.868	542.832	571.332	571.284	.
China	2.585.664	2.983.992	.	.	.	.	.	.	.
India	483.468	521.376	.	523.488	560.100	561.276	600.972	631.788	.
Indonesia	483.468	521.376	.	.	.	.	.	.	.
Canada[a]	32.844	27.960	33.708	34.620	35.376	38.280	35.676	.	.
Kazakhstan	181.572	209.340	.	.	.	.	.	.	.
Poland[b]	84.240	77.916	76.596	76.212	79.524	76.452	73.008	72.228	70.584
South Africa	252.252	249.120	254.160	252.672	258.192	253.704	.	.	.
Russia	245.568	230.316	245.088	257.328	276.504	274.800	286.428	298.788	311.688
United States[c]	1.063.044	975.156	983.724	993.936	922.440	893.436	907.224	813.720	670.104
Ukraine[c]	59.268	54.816	54.444	61.752	64.692	63.312	44.688	28.224	.
Vietnam	39.600	43.752	44.676	44.496	42.096	.	.	.	.

a including sub-bituminous coal, b Excluding briquettes, sub-bituminous coal and similar solid fuels, *c* including brown coal

Not all countries make their recent commercial data available, therefore, there is no information in table 3 about production of such big players as China, Indonesia and Kazakhstan. However, the USA,

IOP Conf. Series: Materials Science and Engineering **268** (2017) 012002 doi:10.1088/1757-899X/268/1/012002

Russia and Australia publish their production. Global hard coal production in 2014 was over 7.2 billion Mg. Poland with production of c. 70 million Mg in 2016 is no longer a coal power. Table 3 has been included mainly as an indication from which country Poland could import coal after its own hard coal deposits have been depleted.

Table 4 shows global hard coal reserves broken down by continents. They amount to c. 411.3 billion Mg. The USA has the largest share in those reserves (108.95 billion Mg).

Table 4. Summary of hard coal balance reserves on individual continents.

No.	Continent	Reserves (million Mg)	Largest share in continent reserves
1.	Europe and Eurasia	102.042	Russia (49 088)
2.	Asia and Oceania	155.809	China (62 200)
3.	South and Central America	6.964	Columbia (6 434)
4.	North America	113.281	USA (108 950)
5.	Africa and Middle East	33.225	South Africa (30 408)
6.	WORLD TOTAL	411.321	USA (108 950)

2.2. Brown coal

Global production of brown coal in 2014 amounted to 810,467 thousand Mg. The largest producer of brown coal is Germany, where in 2015 production amounted to 178,068 thousand Mg. Countries with similar brown coal production of between 63 million Mg and 73 million Mg include: USA, Turkey, Poland and Russia. Brown coal production data between 2008 - 2015 are presented in table 5 [5].

A subtracting in Poland brown coal is usually transported directly to a power plant. Electricity production based on brown coal accounts for c. 31% of total production.

Balance reserves of brown coal in Poland are estimated at c. 23 billion Mg. Reserves broken down by exploration and development categories are shown in table 6. A substantial part of balance reserves (c. 3,690 million Mg) is deposited in so-called Poznań fault and cannot be mined due to protests of the local population.

Table 5. World's brown coal production between 2008 - 2015 in thousand Mg. Source: GUS (Central Statistical Office).

COUNTRY		2008	2009	2010	2011	2012	2013	2014	2015
World		990.528	856.142	840.249	890.675	887.187	834.669	810.467	.
Australia		69.907	72.037	72.090	71.000	71.350	62.845	60.661	.
Bosnia		5.518	5.628	5.618	6.289	5.836	5.711	6.110	.
Brazil		2.229	2.049	2.095	2.136	3.170	3.645	.	.
Bulgaria		28.740	27.180	29.304	37.080	33.420	28.620	.	35.856
Czech Republic		47.532	45.408	43.656	46.620	43.536	40.392	38.220	38.100
Estonia		16.044	14.928	17.892	18.732	18.420	20.508	20.436	19.716
Greece		62.460	61.824	53.628	56.796	61.740	52.572	48.024	45.096
India		32.124	35.724	37.733	40.380	46.524	44.760	46.164	44.448
Indonesia		.	.	.	.	.	.	.	.
Canada	*a*	34.908	34.980	34.188	32.496	31.188	30.624	32.076	.
Kazakhstan		4.728	.	.	.	.	.	.	.
Mecedonia		.	.	6.724	8.209	7.310	6.686	7.128	.
Mongolia		5.903	7.349	5.723	6.938	5.936	6.330	6.330	.
Germany		175.308	169.680	169.404	176.508	185.436	182.700	178.512	178.068
Poland		59.668	57.108	56.510	62.841	64.280	65.849	63.877	63.060
Russia		80.496	68.136	76.620	76.524	77.832	72.984	69.408	73.248
Romania		31.956	28.512	28.668	33.396	32.088	22.896	.	24.204
Serbia		38.520	38.280	37.800	40.812	38.028	40.092	29.940	37.656
United States		68.659	65.751	70.970	73.574	71.602	70.061	72.110	.
Thailand		17.976	16.356	.	21.324	18.072	18.348	17.988	.
Turkey		86.076	82.164	74.760	79.020	74.652	63.048	66.936	51.444
Hungary		9.404	8.986	9.113	9.555	9.290	9.558	.	.

a including sub-bituminous coal

Table 6. Brown coal - million Mg (Source: own study based on data of Polish Geological Institute).

Item	Number of deposits	Geological reserves				Industrial reserves
		balance			off-balance	
		Total	$A+B+C_1$	C_2+D		
TOTAL RESERVES	91	23.516.19	6.067.91	17.448.28	3.522.44	1.122.670
including - developed deposits of resources						
Deposits of operating plants	9	1.418.70	1.406.82	11.89	48.31	1.112.23
including - undeveloped deposits of resources						
Total	74	22.081.18	4.645.43	17.435.75	3.447.62	16.83
1. Deposits explored in detail	35	5.838.65	4645.43	1.193.23	872.64	16.83
2. Deposits explored initially *	39	16.242.52	0.00	16.242.52	2.574.98	-
including - no longer mined deposits						
Production stopped	8	16.30	15.66	0.64	26.51	-

* including resources of deposits in so-called Poznań fault of 3,690 million Mg

Developed balance reserves amount to 1.418 million Mg. This accounts for only 6% of all balance reserves. Coal from those reserves is mined by 5 mining plants, i.e.: Bełchatów, Turów, Adamów, Konin and Sieniawa [3].

Global reserves of brown coal are estimated according to World Energy Council (WEC) at c. 195 bn Mg. In Poland only developed deposits can be considered for mining, i.e. c. 1,418 million Mg of reserves, which at the current average coal production of about 60 million metric tons gives about 24 years. Therefore the Polish energy industry will be able to use brown coal until 2040 at the current production level.

After 2040 brown coal must be substituted by other fuels, e.g. nuclear energy. The same applies to hard coal after 2070.

2.3. Natural gas

Industrial reserves of natural gas in Poland are estimated at c. 54.913 million m^3. Gas deposits can be found in the following regions of Poland: Baltic Sea, Carpathian Mountains, Lowlands and Foothills Developed and undeveloped gas deposits amount to about 51 billion m^3, including c. 3.9 billion m^3 of gas in undeveloped deposits. The balance reserves of the developed deposits amount to c. 101 billion m^3 of gas [3]. Details of natural gas reserves in Poland are presented in table 7.

Natural gas consumption in Poland in 2016 was about 15 billion m^3. Therefore, if the balance reserves were used completely, they would last for seven years at the most. Natural gas production from domestic deposits is about 5.2 billion m^3 [3]. In such a situation gas reserves would last for about 20 years. However, it is impossible to extract completely gas from the deposit, so the period of its exploitation will be definitely shorter. Natural gas extraction from individual deposits in 2015 is presented in table 8. The data presented in Table 8 show that the Lowland and Foothills deposits were exploited most intensively.

Generally, it can be stated that natural gas deposits in Poland are insufficient for the needs of the Polish economy. A lot of hope was put on shale gas, widely present in Poland - figure 3. Reserves of this gas, initially considered as considerable, shrunk as they were identified in detail [6]. Current reserves of this gas are estimated at 346 to 762 billion m^3 (according to PIG).

Natural gas production in Poland compared to selected countries is presented in table 9. Global production in 2015 was 138,479 PJ. The largest share in this production had the USA (29,426 PJ) and Russia (24,386 PJ) [5]. To obtain approximate values expressed in billion $m^{3,}$ multiplythe values in PJ by the conversion factor dependent on gas calorific value, i.e. from 36 to 42. Then production in the USA will be c. 1,059,336 billion m^3, and production in Russia will be: 877,896 million m^3 (Note! the lower value has been used).

Figure 3. Map of shale gas occurrence in Poland
(Source: http://www.gazownictwopolskie.pl/gaz-z-%C5%82upk%C3%B3w/polskie-perspektywy)

Table 7. Natural gas - million m^3 (Source: own study based on data of Polish Geological Institute).

Item	Number of deposits	Recoverable resources				Industrial reserves
			balance		off-balance	
		Total	A+B	C		
TOTAL RESERVES - total	292	122.820.02	75.746.96	47.073.06	2.220.70	54.913.68
from oil andcondensate deposits		27.168.97	10.480.93	16.688.04	654.52	10.642.95
from gas deposits		88.929.48	58.544.46	30.385.02	1.566.18	43.691.34
from underground gas storage facilities		6.721.57	6.721.57	-	-	579.39
including - developed deposits of resources						
Total-	207	101.679.10	69.965.50	31.713.60	663.04	51.006.78
		14.695.13	6.259.36	8.435.77	651.04	6.821.06
		80.262.40	56.984.57	23.277.83	12.00	43.606.33
		6.721.57	6.721.57	-	-	579.39
Baltic Sea (off shore)	2	542.26	109.46	432.80	-	525.81
		542.26	109.46	432.80	-	525.81
		-	-	-	-	-
		-	-	-	-	-
Carpathian Mountains	28	1.115.84	682.56	433.28	10.77	386.72
		115.25	96.83	18.42	1.04	3.73
		879.09	464.23	414.86	9.73	261.49
		121.50	121.50	-	-	121.50
Lowlands	94	68.444.57	50.381.47	18.063.10	650.00	41.460.02
		12.485.17	4.558.05	7.927.12	650.00	5.404.08
		49.874.93	39.738.95	10.135.98	-	36.055.94
		6.084.47	6.084.47	-	-	-
Foothills	83	31.576.43	18.792.01	12.784.42	2.27	8.634.23
		1.552.45	1.495.02	57.43		887.44
		29.508.38	16.781.39	12.726.98	2.27	728.890.00
		515.60	515.60			457.89
including - undeveloped deposits of resources						
Total-	53	20.775.83	5.755.85	15.019.98	1.421.68	3.889.11
		12.371.96	4.221.57	8.150.39	1.93	3.820.87
		8.403.87	1.534.28	6.869.59	1.419.75	68.24

Table 8. Natural gas production - million m^3
(Source: own study based on data of Polish Geological Institute).

Item	Total	From documented balance reserves			off-balance
		Total	A+B	C	
Total	5,213.52	5,212.04	4,226.02	986.02	1.48
from gas deposits	4,455.91	4,455.29	3,705.87	749.42	0.62
from crude oil deposits	367.14	366.28	129.70	236.58	0.86
from condensate deposits	390.47	390.47	390.45	0.02	-
Baltic Sea (off shore)	18.24	18.24	14.69	3.55	-
from gas deposits	0.00	0.00	-	-	-
from crude oil deposits	18.24	18.24	14.69	3.55	-
from condensate deposits	0.00	0.00	-	-	-
Carpathian Mountains	32.84	31.99	27.06	4.93	0.85
from gas deposits	29.61	29.10	25.17	3.93	0.51
from crude oil deposits	3.23	2.89	1.89	1.00	0.34
from condensate deposits	0.00	0.00	-	-	-
Lowlands	3,697.13	3,696.61	3,212.92	483.69	0.52
from gas deposits	3,019.09	3,019.09	2,766.94	252.15	-
from crude oil deposits	287.59	287.07	55.53	231.54	0.52
from condensate deposits	390.45	390.45	390.45	-	-
Foothills	1,465.31	1,465.20	971.35	493.85	0.11
from gas deposits	1,407.21	1,407.10	913.76	493.34	0.11
from crude oil deposits	58.08	58.08	57.59	0.49	-
from condensate deposits	0.02	0.02	-	0.02	-

Table 9. World's natural gas production between 2008 – 2015
Source: GUS (Central Statistical Office).

in PJ (to obtain approximate a value in billion m^3 multiplya given values by conversion factor from 36 to 42)

COUNTRY	2008	2009	2010	2011	2012	2013	2014	2015
World	121.361	117.882	126.332	126.332	132.003	135.455	136.264	138.479
Algeria	3.433	3.299	3.348	3.348	3.374	3.206	3.266	3.257
Saudi Arabia	2.840	2.855	2.787	2.787	3.081	3.114	3.235	3.300
Argentina	1.821	1.729	1.645	1.645	1.589	1.507	1.513	1.554
Australia	1.698	1.722	2.069	2.069	2.123	2.426	2.462	2.594
Azerbaijan	634	635	651	651	691	715	753	762
Bangladesh	655	713	774	774	818	882	905	960
China	3.126	3.320	3.729	3.729	4.306	4.705	5.067	5.214
Egypt	2.366	2.359	2.159	2.159	2.056	2.099	2.146	1.955
Netherlands	2.786	2.624	2.951	2.951	2.672	2.873	2.333	1.776
India	1.223	1.792	1.999	1.999	1.551	1.351	1.279	1.235
Indonesia	2.991	3.120	3.480	3.480	3.132	3.105	3.056	2.952
Iran	5.025	5.408	5.662	5.662	6.154	6.169	6.873	7.237
Canada	6.746	6.298	6.160	6.160	6.033	6.064	6.400	6.378
Qatar	3.242	3.701	4.992	4.992	5.977	6.759	6.624	6.789
Kazakhstan	1.000	1.099	1.145	1.145	1.329	1.428	1.455	1.530
Libya	651	604	639	639	464	486	473	444
Malaysia	2.390	2.175	2.373	2.373	2.394	2.707	2.737	2.699
Mexico	1.775	1.787	1.981	1.981	1.897	1.883	1.734	1.627
Nigeria	1.206	882	1.237	1.237	1.566	1.412	1.612	1.660
Norway	4.053	4.219	4.429	4.429	4.688	4.447	4.419	4.773
Oman	1.057	1.095	1.106	1.106	1.219	1.317	1.260	1.298
Pakistan	1.231	1.270	1.255	1.255	1.262	1.234	1.224	1.291
Poland	172	171	172	172	182	178	173	171
Russia	24.879	22.071	25.128	25.128	25.158	26.204	24.083	24.386
United States	21.855	22.317	23.018	23.018	25.976	26.238	28.027	29.426
Thailand	1.071	998	1.150	1.150	1.219	1.321	1.348	1.276
Trinidad and Tobago	1.629	1.683	1.667	1.667	1.597	1.604	1.596	1.503
Turkmenistan	2.671	1.445	1.716	1.716	2.616	2.972	3.033	3.163
Ukraine	837	752	718	718	717	746	699	680
Uzbekistan	2.555	2.321	2.278	2.278	2.384	2.259	2.339	2.227
Venezuela	1.070	1.023	891	891	945	911	914	1.005
Great Britain	2.917	2.500	2.394	2.394	1.630	1.529	1.532	1.660
United Arab Emirates	1.959	1.904	1.933	1.933	2.046	2.057	2.042	2.214

2.4. Crude oil

Crude oil reserves in Poland are very small, therefore, this energy raw material is almost entirely imported, in particular, from Russia. Crude oil reserves in Poland are shown in table 10. As at 31 December 2015 balance reserves of crude oil totalled c. 23 million Mg, and industrial reserves were only c. 14 million Mg. This material is present together with natural gas in the following deposits: Baltic Sea (off shore), Carpathian Mountains, Lowlands and Foothills The reserves in table 10 are shown as balance, off-balance and industrial ones, and developed and undeveloped, and deposits exploitation of which has stopped.

Crude oil production in the world and Poland between 2010 - 2015 is presented in table 11. The data in table 11 show that, similarly to natural gas production, the biggest crude oil producers in 2015 were: The USA – 640,699 thousand Mg, Russia – 501,857 thousand Mg. Poland with production of 894 thousand Mg occupies the last position in table 11 [5]. Refining capacities of the petrochemical industry in Poland are c. 21 million Mg per annum. The main consumers of crude oil in Poland are PKN ORLEN and LOTOS Gdańsk Refinery. Crude oil is supplied to those companies mainly through the 'Przyjaźń' pipeline and the Northern Port in Gdańsk. In the light of the presented figures it can be stated that Poland does not have crude oil as a raw material, which could ensure its energy security.

Table 10. Crude oil - thousand Mg
(Source: own study based on data of Polish Geological Institute).

Item	Number of deposits	Recoverable resources				Industrial reserves
			balance			
		Total	A+B	C	off-balance	
TOTAL RESERVES - total	86	22.824.28	9.496.80	13.327.48	395.40	14.191.12
Crude oil		21.338.91	9.473.22	11.865.69	395.40	14.131.10
Oil condensate		1.485.37	23.58	1.461.79	-	60.02
including - developed deposits of resources						
Total-		22.260.68	9.378.91	12.881.77	9.76	14.074.24
		20.931.13	9.378.91	11.552.22	9.76	14.074.24
		1.329.55		1.329.55	-	
Baltic Sea		4.654.27	1.059.10	3.595.17		4.518.44
		4.654.27	1.059.10	3.595.17	-	4.518.44
Carpathian Mountains		547.18	441.18	106.00	6.55	47.35
		547.18	441.18	106.00	6.55	47.35
Lowlands		16.682.91	7.652.79	9.030.12	3.21	9.460.43
		15.353.36	7.652.79	7.700.57	3.21	9.400.41
		1.329.55	-	1.329.55	-	60.02
Foothills		376.32	225.84	150.48		48.02
		376.32	225.84	150.48	-	48.02
including - undeveloped deposits of resources						
Total-		507.03	108.29	398.74	329.53	116.50
		363.03	89.29	273.74	329.53	116.50
		144.00	19.00	125.00	-	-
Lowlands		391.10	108.29	282.81	-	116.50
		247.10	89.29	157.81	-	116.50
		144.00	19.00	125.00	-	
Foothills		115.93		115.93	329.53	
		115.93	-	115.93	329.53	
including deposits exploitation of which has stopped						
Total-		56.57	9.60	46.97	56.11	0.38
		44.75	5.02	39.73	56.11	0.38
		11.82	4.58	7.24	-	-
Carpathian Mountains		1.50		1.50	3.75	-
		-	-	-	3.75	
		1.50	-	1.50	-	
Lowlands		50.49	5.02	45.47	1.43	0.38
		44.75	5.02	39.73	1.43	0.38
		5.74	-	5.74	-	-
Foothills		4.58	4.58		50.93	
		-	-	-	50.93	-
		4.58	4.58	-	-	

IOP Conf. Series: Materials Science and Engineering **268** (2017) 012002 doi:10.1088/1757-899X/268/1/012002

Table 11. World's crude oil production between 2008 – 2015 in thousand Mg [4]
Source: GUS (Central Statistical Office).

COUNTRY	2008	2009	2010	2011	2012	2013	2014	2015
World	3.728.306	3.634.622	3.689.826	3.718.712	3.799.594	3.783.850	3.857.747	.
incl:								
Algeria	61.048	54.736	54.048	52.976	55.776	54.617	54.275	.
Angola	93.790	85.582	84.588	79.670	83.932	83.820	81.349	86.807
Saudi Arabia	460.145	407.938	406.997	464.098	487.968	480.355	484.116	508.026
Argentina	32.352	31.138	30.690	29.047	28.766	26.996	26.954	27.194
Australia	23.742	23.032	23.729	20.585	20.778	17.262	18.145	16.457
Azerbaijan	44.026	50.539	51.035	45.487	42.985	43.084	41.880	41.311
Brazil	92.376	99.500	104.801	107.436	105.398	103.147	114.910	124.178
China	190.024	189.606	203.832	203.646	206.998	208.202	209.450	.
Egypt	33.264	33.150	35.928	35.376	36.269	35.338	.	.
Ecuador	26.196	24.256	24.890	26.089	26.326	27.445	29.006	28.320
Equatorial Guinea	17.397	15.406	13.629	13.885	14.434	.	.	.
India	33.972	33.229	36.704	38.237	37.990	37.678	37.543	37.238
Indonesia	48.312	46.700	46.003	44.675	42.592	40.717	39.106	39.046
Iraq	.	114.761	116.971	130.330	145.357	146.275	152.795	172.141
Iran	202.368	176.638	175.976	180.684	189.846	177.575	154.794	156.510
Canada	126.588	125.634	133.879	141.761	153.042	163.595	175.207	178.612
Qatar	41.128	35.677	35.683	35.692	35.797	35.226	34.507	31.915
Kazakhstan	66.584	64.292	.	.	.	.	67.924	66.514
Columbia	30.136	34.556	40.468	47.140	.	.	.	.
Kuwait	134.926	113.860	116.401	133.931	150.269	147.108	144.445	.
Libya	83.368	70.968	71.584	-	70.015	47.728	.	.
Malaysia	33.048	31.427	30.254	27.346	27.864	27.214	.	.
Mexico	145.524	135.246	133.916	132.575	132.464	131.126	126.275	117.851
Nigeria	103.740	104.370	120.083	116.558	114.864	107.975	106.532	103.886
Norway	99.288	98.255	91.316	85.664	78.299	74.382	76.150	77.611
Oman	.	40.459	43.051	44.128	45.865	46.906	46.978	48.847
Poland*	755	687	686	618	678	961	949	894
Russia	471.426	489.450	505.327	509.441	495.199	498.563	500.476	501.857
1	*2*	*3*	*4*	*5*	*6*	*7*	*8*	*9*
United States	337.460	361.228	376.576	391.656	444.247	507.360	591.200	640.699
Sudan	23.098	23.738	23.104	22.632	5.079	.	.	.
Syria	18.647	18.325	19.785	18.033	8.253	.	.	.
Venezuela	163.288	149.347	144.752	147.446	146.644	145.348	139.924	138.888
Great Britain	65.497	62.820	58.046	48.571	42.053	38.456	37.475	42.826
United Arab Emirates	124.914	107.536	111.650	123.204	128.024	146.456	145.928	.

* Petroleum oils and oils obtained from bituminous minerals, crude

3. Renewable energy sources (RES)

Renewable energy sources include water, sun, water, biomass and biogas energy. These sources can be converted into electricity. Renewable energy includes also thermal sources, which are present in small quantities in Poland. They are of insignificant importance for energy security.

The EU energy policy, which should be followed by Poland, provides for systematic increase of the share of renewable energy in the energy mix of the Member States. Installed power in the Polish energy industry from renewable energy sources between 2005 and 2016 is shown in table 12. The data presented in table 12 show that to 2016 included 8.4 GW from RES were installed [7]. The most energy was obtained during that period from wind (58%), water (16%), biomass (22%), biogas (4%) and sun (0,03%).

Table 12. Power installed in the Polish energy industry obtained from renewable energy sources between 2005 – 2016 (Source: Own study based on the Energy Regulatory Office (URE) data)

Year	RES installation type - Installed power [MW]						
	biogas	biomass	solar power	wind power	water power	Total	increase y/y
2005	31.972	189.790	-	83.280	852.495	1.157.537	
2006	36.760	238.790	-	152.560	934.031	1.362.141	204.604
2007	45.699	255.390	-	287.909	934.779	1.523.777	161.636
2008	54.615	231.990	-	451.090	940.576	1.678.271	154.494
2009	70.888	252.490	0.001	724.657	945.210	1.993.246	314.975
2010	82.884	356.190	0.033	1.180.272	937.044	2.556.423	563.177
2011	103.487	409.680	1.125	1.616.361	951.390	3.082.043	525.620
2012	131.247	820.700	1.290	2.496.748	966.103	4.416.088	1.334.045
2013	162.241	986.873	1.901	3.389.541	970.128	5.510.684	1.094.596
2014	188.549	1.008.245	21.004	3.833.832	977.007	6.028.637	517.953
2015	212.497	1.122.670	71.031	4.582.036	981.799	6.970.033	941.396
2016	233.967	1.281.065	99.098	5.807.416	993.995	8.415.541	1.445.508

4. Nuclear Energy

4.1. Nuclear power throughout the world

The information material developed by the Nuclear Energy Department of the Ministry of Energy in March 2017 [8] describes 449 operating nuclear reactors in 30 countries. In 2016 7 new reactors were put into service. Next 60 blocks is under construction and over 160 are planned. As a result global energy production capacities in nuclear power plants increased by about 9 GWe and at the end of 2015 their total power reached 392.23 GWe. Electricity production in nuclear power plants also increased and in 2015 was 2,441 TWh. The share of nuclear energy in total electricity production has been at a similar level for four years, and in 2015 it increased by hefty 11.5%. Apart from reactors generating electricity, there are also 228 research reactors used for scientific and educational research.

Table 13 presents the number of reactors, installed power and electricity production in those reactors. After a short period of investment stagnation in the nuclear energy industry, the boom in the nuclear energy sector can be currently observed.

Table 13. Energy reactors and world's electricity production in EJ [8].

	Number of reactors	Net installed power (GWe)	Energy production (TWh)	Share (%)
2008	439	371.56	2597	15.0
2009	437	370.7	2558	14.0
2010	441	375.28	2630	13.8
2011	435	368.92	2518	13.5
2012	437	373.26	2346	11.0
2013	434	371.79	2359	11.0
2014	438	376.34	2410	11.0
2015	441	382.86	2441	11.5
2016	448	391.23		
2017	449	392.23		

The number and power of active reactors in the nuclear energy sector is presented in figure 6. Decisions on the construction of new power units or new nuclear power plants are currently taken in developing countries, in particular in South-East Asia (China, India, South Korea). Electricity production in nuclear power plants is shown in figure 4 [8]. Noticeable decrease of production is made up in 2014 and 2015.

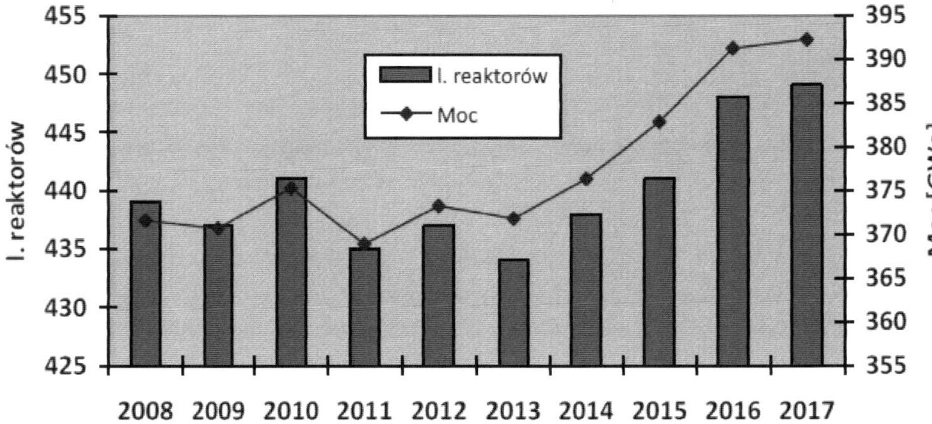

Figure 4. Number and power of world's active power reactors [8].

4.2. Nuclear power in Poland

After long discussions devoted to threats resulting from a potential breakdown of a nuclear power plant, social acceptance of construction of the first nuclear power plant has been obtained. Its opening is foreseen for 2026 and included in 'The assumptions of the Polish energy policy until 2030'. Nuclear power was to gradually substitute coal in the energy mix. Time necessary to construct a nuclear power plant is estimated at about 15 years. Therefore, it is impossible to construct such a power plant before 2030, all the more so since public feeling is also changing against nuclear power. Meanwhile, domestic hard and brown coal deposits are running out and not later than in the years 2050 - 2070 [3] they will have to be substituted with other primary energy source or imported. Public concern is unjustified, because around the Polish borders nuclear power plant already exist. Therefore, any breakdown of those power plants threatens also Poland. The public should be therefore convinced of such solutions.

5. Production and consumption of electricity in Poland

In 'The assumptions of the Polish energy policy' [2] it is assumed that the installed power in the National Power System in 2020 should be around 44 GW. It means increase of installed power by 7 GW. However, in the same document the same amount of power is assumed to be phased out by 2020 due to its wear and necessary modernisation. It is estimated that in the following years 15 GW of new production capacities should be created, which seems to be a difficult task. Current and until 2050 demand for electricity shows that extension of power infrastructure is a must not to experience 'Black point'. Along with the issue of reconstructing production capacities in the power industry, the issue of supplying of energy raw materials to power plants must be solved. Currently the power industry, as is well known, is based on hard and brown coal. These materials will be depleted from the domestic reserves between 2040 and 2070. How to ensure energy security in Poland, whether to rely on coal, even imported, or whether to focus on other, more ecological, energy sources.

Table 14 presents production and consumption of electricity in the years 2015 - 2016. The data in table 14 shows that domestic electricity consumption in 2016 was 161,438 (in GWh) and was higher by 1.97% than in 2015. Total production was higher by 0.53%, and similarly, production in utility hydro power plants was higher by 6.1%. Considerable increase of electricity production was reported in gas-fired power plants (by 37.77%) and renewable energy sources (by 100.03%), including wind farms by 15.76%.

On the other hand, electricity production based on hard coal decreased by 0.65% and on brown coal by 4.41% [9]. Could these data show the withdrawal from coal-based power?

Table 14. Structure of electricity production in domestic power plants, volume of electricity exchange with foreign countries and domestic electricity consumption - monthly volumes and from the beginning of year - gross values (Source: Polskie Sieci Elektroenergetyczne – monthly reports).

Item	December			Cumulatively from January to December		
	2015 (GWh)	2016 (GWh)	Dynamics ((b-a)/a*100) (%)	2015 (GWh)	2016 (GWh)	Dynamics ((e-d)/d*100) (%)
Total production	14.464	15.218	5.22	161.772	162.626	0.53
Utility power plants	11.859	12.559	5.9	141.901	140.727	-0.83
Water utility power plant	229	272	18.64	2.261	2.399	6.1
Thermal utility power plant	11.630	12.287	5.65	139.640	138.328	-0.94
hard coal	6.972	7.381	5.87	81.883	81.348	-0.65
brown coal	4.166	4.375	5.02	53.564	51.204	-4.41
gas	493	532	7.89	4.193	5.776	37.77
Other renewable sources power plants	6	12	100	73	146	100.03
Wind farms	1.612	1.599	-0.78	10.041	11.623	15.76
Industrial power plants	986	1.047	6.2	9.757	10.130	3.82
Foreign exchange balance	-334	-243	-27.26	-334	1.999	-
Domestic electricity consumption	14.129	14.975	5.99	161.438	164.625	1.97

Table 15 present perspective demand for electricity included in 'The assumptions of the Polish energy policy until 2050' [2]. In the assumptions substantial volume of electricity production is based on nuclear power, whose presence in the Polish energy mix at that time is doubtful. If such energy is to be substituted by coal, coal will have to be imported.

Table 15. Electricity production estimate broken down by fuel (TWh) in Poland until 2050
Source: Schedule to 'The Polish energy policy until 2050' (draft).

Energy source	2010	2015	2020	2030	2035	2040	2045	2050
Hard coal	87.9	72.5	76.9	75.9	84.4	88.8	82.3	74.5
Brown coal	48.6	58.4	53.8	49.6	11.1	11.3	10.7	10.3
Natural gas	6.8	5.8	11.8	11.9	18.4	17.5	23.3	20.4
RES	11.6	20.6	34.0	36.9	61.1	65.1	67.5	73.2
Nuclear energy	0.0	0.0	0.0	11.8	45.1	45.4	44.2	43.2
Other	2.6	1.4	1.4	1.4	1.4	1.4	1.4	1.4
Total	157.5	158.7	177.9	187.5	221.5	229.5	229.4	223.0

6. Polish energy security indicators

An indicator determining national energy security is Stirling indicator (formula 1) [10],which determines a degree of diversification of energy supply:

$$d_N = -\sum_{j=1}^{m} u_j \ln u_j \tag{1}$$

where: u_j – share of jth energy carrier in national energy supply structure, m – number of energy carriers, ln – natural logarithm.

Even structure of energy supplied to the domestic market is the most favourable. In Poland the Stirling indicator is about 1.0 and differ from the EU level (1.4), which is caused by a considerable share of domestic solid fuels (coal) in the Polish energy balance. The indicator can be improved by diversification of energy sources supplies. Not more than 30% can be imported from one direction (gas 85%, crude oil 95% from Russia).

The second parameter determining the condition of energy security is country's energy self-sufficiency [10]. It is defined as the ratio of volume of domestic fuel acquisition to primary energy consumption:

$$W_s = \frac{P \cdot 100\%}{Z_k} \tag{2}$$

where: P – fuel production in a given year, Z_k – domestic global consumption equal to the sum of individual fuels supplied to the domestic market less domestic reserves balance.

This indicator is currently about 85% and decreases as own resource diminishes.

Table 16. Global documented industrial reserves of conventional energy raw materials and their sufficiency

Materials	Reserves (million toe)	Reserves structure (%)	Reserves consumption (million toe)	Use structure (%)	Reserves sufficiency (years)	
					world	Poland
Hard/brown coal	469.298	59.6	2.957.0	31.7	158	48 24
Crude oil	159.644	20.3	3.861.3	41.4	41	1
Natural gas	158,815	20.1	2.512.2	26.9	63	7
Total	**787.757**	**100.0**	**9.330.5**	**100.0**	**84**	-

Note: 1 toe = 41.868 GJ

Calculated per m^3 global natural gas reserves as at 2009 amounted to 187.47 trillion m^3. In the following years, despite current production, global natural gas reserves increased due to new discoveries. The situation is similar as far as crude oil is concerned - global reserves of this material according to the situation as at 2013 amounted to 234 billion metric tons. The richest deposits are in Saudi Arabia (36.5 bn t), Iran (20.8 bn t), Iraq (19.3 bn t), Venezuela (46.3 bn t), Kuwait (14.0 bn t) and Canada (28.2 bn t).

7. Summary

The article describes all available energy sources, which can be used to generate electricity, considered to be the cleanest form of energy. Such energy sources include: hard coal, brown coal, natural gas, crude oil, nuclear power and renewable energy sources (RES). Renewable energy sources include, in particular: water, solar, wind, biomass and biogas power.

All the mentioned energy sources, also called primary energy sources, has been described in detail in the context of economic usefulness and analysed in the context of sufficiency for social needs defined as National Energy Security. Individual types of primary energy sources, belonging to fossil fuels, have been analysed in terms of their reserves and recoverability in Poland and the world.

The Polish power industry is currently based on hard and brown coal, which generate c. 81% of total electricity - hard coal accounts for c. 50% from and brown coal for c. 31%. The above are followed by: natural gas (c. 3%) and water power (c. 3%). Water power is recognised as Renewable

Energy Sources. Other renewable energy sources develop according to the guidelines of the European Union and in 2020 they will reach about 15% - it is an estimated share in our energy mix [1]. There are not many good conditions for development of RES in Poland. Apart from the mentioned water power, solar and wind power have the largest share in RES. As far as fossil fuels are concerned, hard coal, at the current level of development and production (c. 70 million Mg/year), will last for about 50 years, i.e. to 2070. Brown coal from currently developed deposits, with annual production of c. 60 million Mg will last for about 30 years, i.e. to 2050. Natural gas, with deposits of about 100-120 billion m^3 and current production of c. 5 billion m^3, will last for about 20 years. However, natural gas consumption in Poland amounts currently to 15 billion m^3 and is likely to increase - therefore Poland is will have to import. When importing it is necessary to diversify gas supplies. The LNG terminal in Świnoujście with annual capacity of 5 billion m^3 of liquefied gas constitutes an important, but not sufficient, support of the national energy security. It is a pity that the project of 2001, which provided for connection of the Norwegian shelf and Poland in Niechorze with a pipeline with annual capacity of 10 billion m^3 of gas, where about 5 billion m^3 were intended for Poland. Currently Poland still has to import gas from the east through the existing pipelines.

Finally, crude oil scarce in Poland - 20-25 million metric tons. At current annual production levels of c. 1 million Mg, it will last for 20-25 years. However, annual needs relating to oil refining in Polish refineries are at a level of c. 20 million Mg. Crude oil is delivered to Poland through the 'Przyjaźń' pipeline from Russia and by tankers to the Northern Port in Gdańsk from other directions. So far such transport system has worked correctly, however, there is a project providing for crude oil transport through a new pipeline, 'Golden Gate', from the Caspian Sea, via Ukraine to Poland.
Assuming that it will be possible to extract the hard and brown coal deposits completely - the country's energy security is ensured until about 2050. If, however, coal will be gradually withdrawn from the Polish energy mix, it is urgent now to seek for other solutions, including nuclear power.

8. Conclusions

- Hard and brown coal reserves are running out in the developed deposits.
- Despite large reserves of brown coal are present in Poland - it should be assumed that they will not be exploited due to opposition of the local population.
- Hard coal reserves, although still quite rich, may not be fully exploited due to increasing natural hazards, such as methane, crumps and rockmass temperature. Also in this case Silesian community opposes to building new coal mines (e.g. near Pszczyna).
- The EU guidelines on environmental protection do not promote development of coal-based power generation.
- In the search for new energy sources it is necessary to start to construct nuclear power plants to make up for the losses of coal-based power generation.
- An alternative to coal power plants can be also gas power plants, especially in the light of the reports of the Polish Geological Institute (PIG) on shale gas deposits (deposits of 346 to 768 billion m^3).
- Nuclear power as well gas power plants (if shale gas is not available) are based on imported primary energy sources.
- Decrease of hard and brown coal production to extend the period of its production and introduction of other energy sources (gas, nuclear power) should also be considered. A source of uranium supplies to Polish nuclear power plants is equally important. Considerable deposits of uranium can be found in Canada.

References

[1] Energy Law Act of 10 April 1997 as amended (J.L. 153.1504)
[2] Draft of Polish energy policy until 2050 (in polish: Projekt Polityki energetycznej Polski do 2050 roku), http://bip.me.gov.pl/node/24670, access of 11 May 2017.
[3] Balance of fossil fuel reserves in Poland as at 31 December 2015, Polish Geological Institute,

Polish Reserach Institute Warszawa 2016.

[4] Sobczyk E J, Kicki J, Jarosz J, Kowalczyk I and Stachurski K 2016 Gospodarka zasobami złóż węgla kamiennego w Polsce w latach 1990–2015 *Zeszyty Naukowe Instytutu Gospodarki Surowcami Mineralnymi i Energią Polskiej Akademii Nauk* **92** pp 37–56.

[5] GUS reports, March 2017 (in polish: Raporty GUS marzec 2017), http://stat.gov.pl/statystyka-miedzynarodowa/porownania-miedzynarodowe/tablice-o-krajach-wedlug-tematow/przemysl-i-budownictwo/, access of 10 May 2017.

[6] Polish shale gas reserves https://www.polskielupki.pl/artykul-gaz-lupkowy-w-polsce/74129/ zasoby-gazu-lupkowego-w-polsce, access of 21 April 2017.

[7] Installed power, reports of URE https://www.ure.gov.pl/pl/rynki-energii/energia-elektryczna/ odnawialne-zrodla-ener/potencjal-krajowy-oze/5753,Moc-zainstalowana-MW.html, access of 10 May 2017.

[8] World's nuclear energy, Information material developed by the Nuclear Energy Department of the Ministry of Energy, March 2017.

[9] Polskie Sieci Elektroenergetyczne – monthly reports http://www.pse.pl/index.php?modul =8&y=2016&m=12&id_rap=212, access of 08 May 2017.

[10] Kaliski M and Staśko D 2003 Rola krajowej infrastruktury paliwowo – surowcowej w kształtowaniu bezpieczeństwa energetycznego Polski *Wiertnictwo, nafta, gaz* **20/1** pp 101-10.

IOP Conf. Series: Materials Science and Engineering **268** (2017) 012003 doi:10.1088/1757-899X/268/1/012003

Trends in coal use – global, EU and Poland

Wojciech Suwała, Artur Wyrwa and Tadeusz Olkuski

AGH University of Science and Technology, the Faculty of Energy and Fuels, Mickiewicza 30, 30-619 Kraków, Poland

olkuski@agh.edu.pl

Abstract That aim of this paper is to compare trends in global, European use of coal with tendencies in Poland, one of heavy coal dependent countries. Polish power generation is unique among OECD countries, the share of both hard coal and lignite in power generation reaches 81% [1]. Climate policy of European Union is to phase out intensive greenhouse gases sectors, thus to transform Polish power generation into less carbon intensive. Although such policy is generally accepted in Poland, the paste and practically proposed regulation that excludes coal generation from capacity mechanisms, is considered as threat to energy security. Coal is the base for generation for one simple reason, abundant in European scale hard coal reserves and significant capacities in lignite. Natural gas reserves allow to supply about 1/3 of consumption, but prices and supplies dependent hitherto on contracts with GAZPROM did not allow to develop significant generation capacities. Renewable resources are limited, there is not much possibilities for hydro, wind and solar. Poland is also one of the countries of poor air quality, traditional coal based space heating systems plus obsolete car fleet generate vast emissions, especially during the winter. Only recently this became top priority of environmental authorities. This situation is subject to transformation, government, managers are aware that the role of coal needs to be decreased, but there are two main questions, the paste of transformation and the future energy mix. The paper attempts to answer the question whether the expected changes in Polish energy mix are comparable or differ from the global and European tendencies.

1. Global view

The major force that determines use of coal is the climate change which was identified and is observed in recent decades, and Intergovernmental Panel on Climate Change declares: "*It is extremely likely that human influence has been the dominant cause of the observed warming since the mid-20th century*" [2]. On the other hand climate on earth was never stable and is continuously changing.

Carbon dioxide is greenhouse gas which concentration rises since the beginning of the industrial era thus claimed to be the main reason for growing earth temperature contributing to climate change. The main sector responsible for greenhouse gases emissions is energy, mainly electricity and heat production and other energy processes (figure 1) and their share is estimated at about 35% of total.

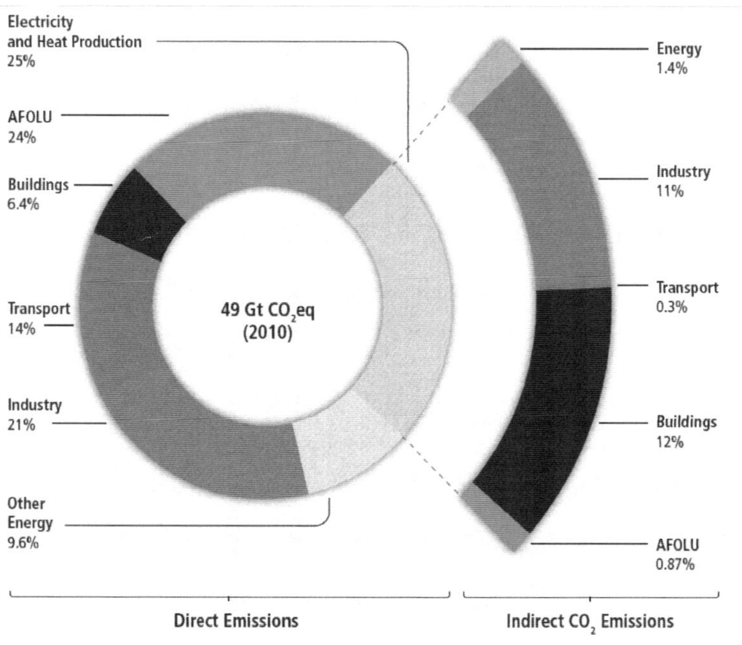

Figure 1. Greenhouse gases emissions by sectors [3].
(AFOLU is agriculture, forestry and other land use sectors)

China is the country of largest carbon emissions, overcoming twice next on the list – USA (table 1). Poland with emissions of 78Mt or less than 0,9% is ranked as 21st .

Table 1. Ranking of the selected countries by 2014 total CO2 emissions (expressed in thousand metric tons of carbon) [4]

Rank	Country	Carbon emissions	Percent
1	CHINA (MAINLAND)	2806634	30.16%
2	UNITED STATES OF AMERICA	1432855	15.40%
3	INDIA	610411	6.56%
4	RUSSIAN FEDERATION	465052	5.00%
5	JAPAN	331074	3.56%
6	GERMANY	196314	2.11%
15	UNITED KINGDOM	114486	1.23%
17	TURKEY	94350	1.01%
18	ITALY (INCLUD. SAN MARINO)	87377	0.94%
19	THAILAND	86232	0.93%
20	FRANCE (INCLUDING MONACO)	82704	0.89%
21	POLAND	77922	0.84%
22	TAIWAN	72013	0.77%
23	KAZAKHSTAN	67716	0.73%
	Total	9306200	100.00%

Following pictures present global perspectives on GHG emissions and power generation, from World Energy Outlook prepared by International Energy Agency in three scenarios: CP - Current Policies, NP - New Policies, and 450 Scenario. The latter assumes radical climate policy not to exceed 450 ppm of carbon dioxide. The New Policies Scenario assumes that countries follow their Nationally Determined Contributions, as declared after COP21 in Paris. However, today this scenario seems not to be attainable due to the declaration of abandoning commitments by USA administration.
It is expected that the energy related carbon dioxide emissions in growing economies will increase due to the economy progress while slightly decreasing in OECD countries (figure 2).

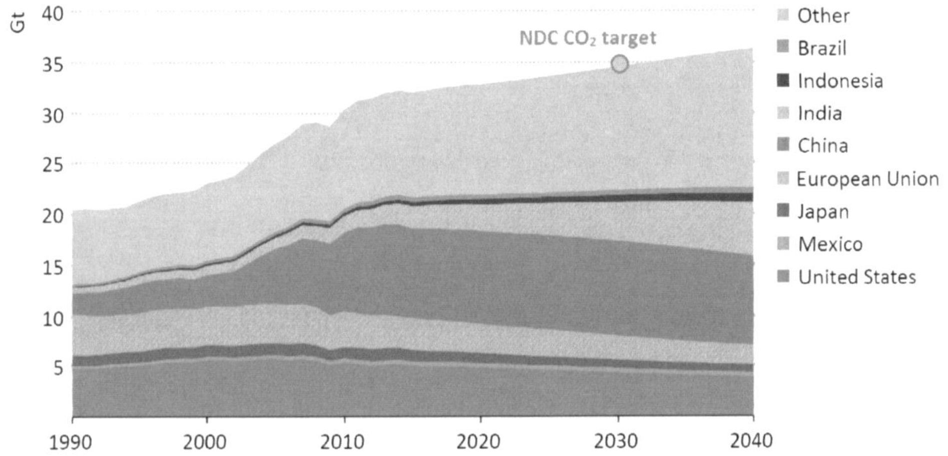

Figure 2. Energy-related CO_2 emissions by region in the New Policies Scenario [5].
(NDC CO2 target - Nationally Determined Contributions related to the COP21 Paris Agreement)

In the NP and CP scenarios energy consumption should rise from present 24 thousands of TWh to by some 25% to 2025 and 65% to 2040 in NP scenario while in CP 30% and 80% respectively (figure 3).

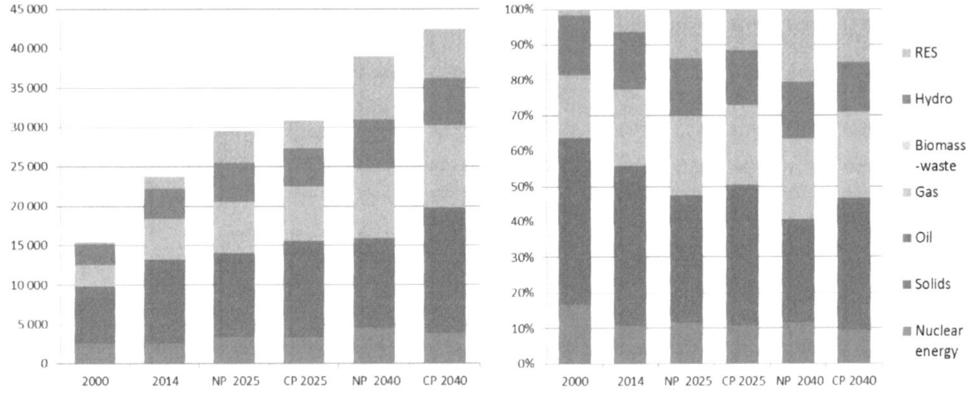

Figure 3. Global electricity production and structure, TWh (based on [5]).
(scenarios: CP Current Policies, NP New Policies)

2. Comparing European and Polish trends and prospects

Present generation of electricity in Poland is dominated by coal both hard and lignite, what was inherited from the previous less environment aware periods and limited access to other fuels. On the other hand such structure guarantees supply security, given experiences even with hard winters.
This structure started to change, other sources gradually gain significant share (figure 4).

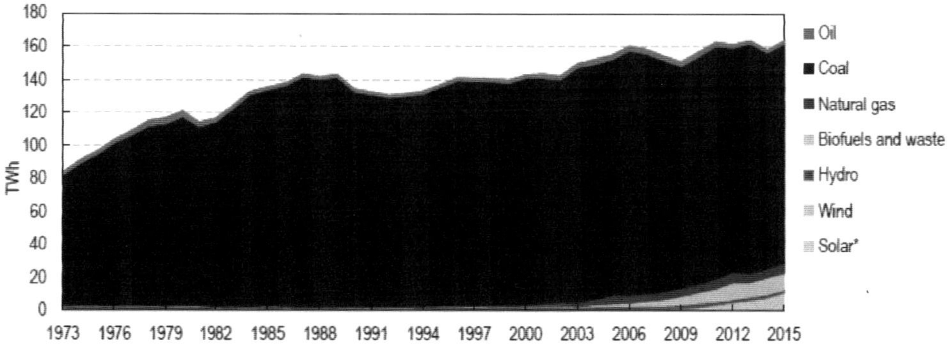

Figure 4. Poland power generation 1973-2015 [TWh] [1].

Later discussion will present prospects of the alternation of this tendency to accelerate GHG emissions abatement.

Trends in GHG emissions in European Union and Poland (data based on [6]) (figure 5) show that the pace for the period 1990 – 2014 was comparable, both EU and Poland reduced emissions by 20-30%. However, Polish emissions decreased significantly in the period 1990 – 2000, being practically at the same level thereafter. Recent 10 years for EU present significant acceleration in emissions reduction, as well as Polish emissions in 5 recent years. The latter should be verified in the next periods.

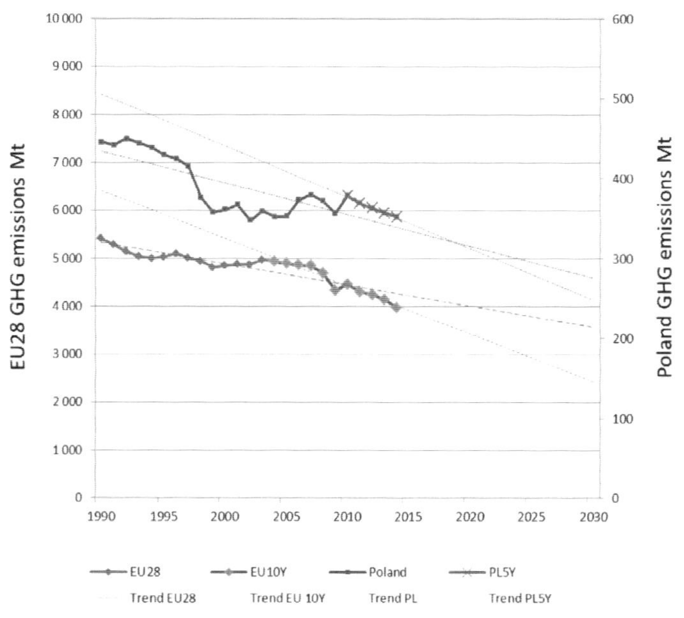

Figure 5. Trends in GHG emissions EU28 and Poland (based on [6]).
(EU 10Y represents last 10 years for EU28, PL5Y last 5 years for Poland)

2.1. Reference energy mix for Poland

In 2016 the European Commission published the latest reference scenario for Poland prepared with the use of the PRIMES model. This has created an opportunity to compare the results presented therein with the results of the TIMES-PL model. PRIMES is an energy-economic model developed by E3MLab of the Polytechnic University in Athens. It covers all EU countries [7]. TIMES-PL is a model of the Polish power system developed at the AGH University of Science and Technology [8]. It belongs to a family of national models developed using the TIMES generator. Both PRIMES and TIMES-PL are used for modeling the development of the power system in the medium and long term. It should be noted that the results of both models are not forecasts of the most probable state of the system in the future, but rather explore possible system development trends for certain determinants and assumptions. This type of research makes it possible to analyze the impact, especially of the various environmental and energy policy instruments, on the development of the energy system. A similar comparative analysis was made for the 2013 EETT2050 Reference Scenario [9]. Nevertheless, since then there has been some changes, especially in regards to the European emissions trading system, for which higher prices of CO_2 emission allowances are foreseen. Another significant change is a greater pessimism towards the commercial use of carbon capture and storage technology due to slow development of this technology even under public support of the European Economic Recovery Program (EEPR). Consequently, CCS technology is only being used since 2040 and to a much lesser extent than it was presented in the previous EETT report. Prices of emission allowances, main energy carriers and electricity demand forecasts used in the Reference Scenario in both models are presented in table 2. The technical and economic parameters characterizing the energy technologies are presented respectively in EETT 2016 for PRIMES and in GIPH [10] for TIMES-PL.

Table 2. Major assumptions on prices in the Reference Scenario for European Union (PRIMES) and Poland (TIMES-PL).

	2015	2020	2030	2040	2050
EUA €/Mg CO_2	8	15	33	50	88
PRIMES:					
Gas $/1000 m3	290	390	445	480	500
Coal in import to EU $/Mg	77	95	102	119	77
Gross Electricity generation [TWh]	162.4	176.2	203.2	219.2	245.3
TIMES-PL:					
Gas $/1000 m3	308	300	287	274	268
Coal [$/Mg]	84	89	102	118	137
Gross Electricity generation [TWh]	163.7	175.5	197.4	223.6	240.5

Figure 6 shows the fuel structure of electricity generation according to the PRIMES model for all 28 EU Member States. As anticipated, more ambitious CO_2 reduction targets under the current EU climate policy are raising the prices of CO_2 allowances and result in gradual displacement of fossil fuels from the EU's energy mix and their substitution by renewable energy sources, mainly wind turbines and solar power plants. Figure 7 shows Poland power generation capacities and its structure.

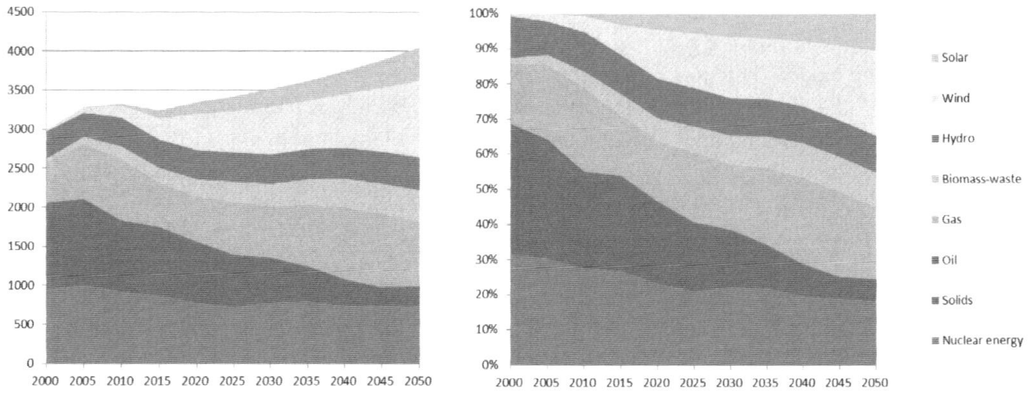

Figure 6. EU28 power generation [TWh] and structure (based on EU Reference ...).

PRIMES results for Poland show a similar trend. Significant decreases in the use of hard coal and lignite. Power units which are planned to be decommissioned in the period after 2030 are not renewed using coal technologies but rather in technologies using natural gas, renewables and nuclear.

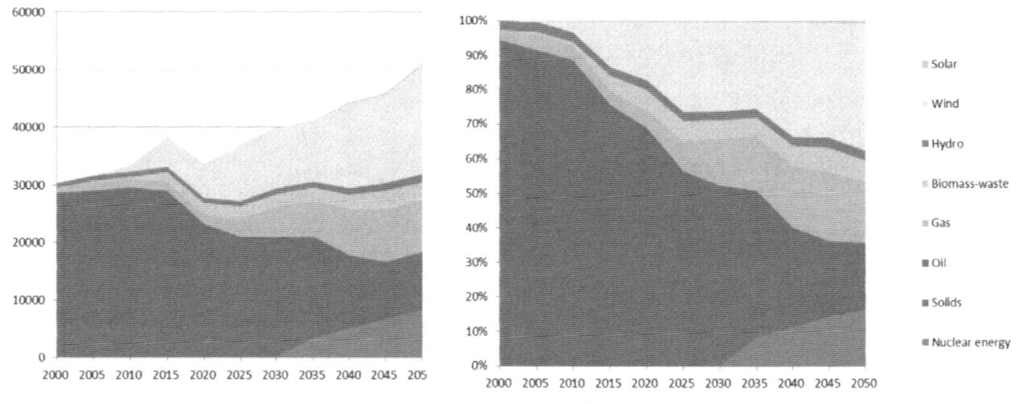

Figure 7. Poland power generation capacities [GW] and structure (based on EU Reference ...).

In case of the TIMES-PL model the assumption about the CCS potential seems to be crucial. With the assumptions that CCS technology will be available after 2035 and that there will be social acceptance for the opening of new coal mines, both hard coal and lignite remain the main fuels used to produce electricity. Nevertheless, the share of solid fuels in electricity production is decreasing to 48% in 2050 (figure 8).

IOP Conf. Series: Materials Science and Engineering **268** (2017) 012003 doi:10.1088/1757-899X/268/1/012003

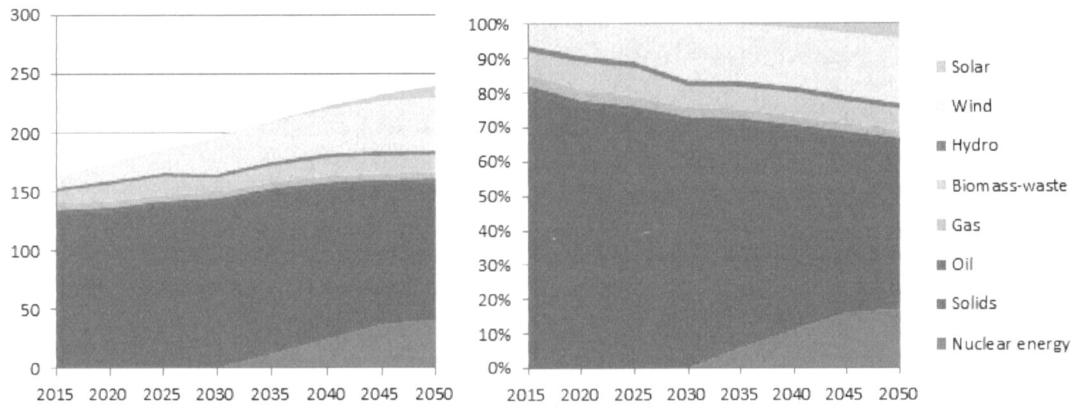

Figure 8. Poland, power generation [TWh] and structure with CCS – TIMES-PL.

The situation is different with the assumption (which is similar to the one in the EETT scenario) that there will be a delay in the possibility of CCS use in the power industry. Figure 9 depicts the TIMES-PL results concerning the fuel structure for electricity generation for the case when electricity production in CCS-equipped power plants is reduced to the values provided in the EETT report.

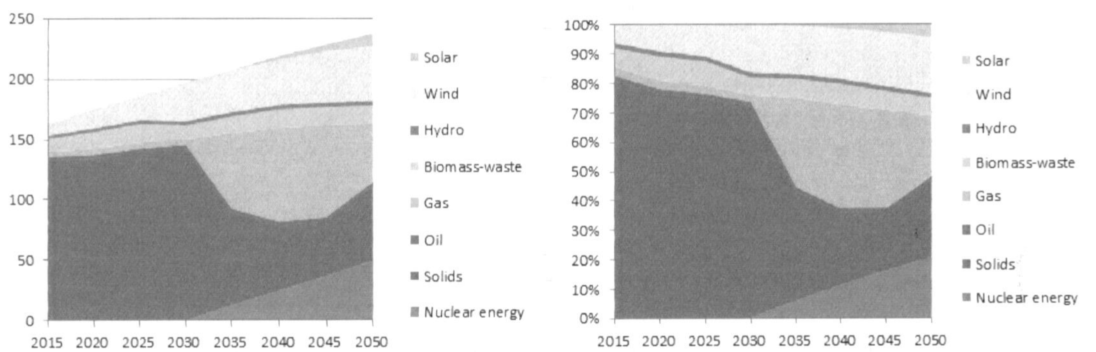

Figure 9. Poland, power generation [TWh] and structure for the case when electricity production in CCS-equipped power plants is reduced to the values provided in the EETT report – TIMES-PL.

It is clear that with such high prices of CO_2 emission allowances and with limitations for the application of CCS technology, the use of coal for power generation becomes unprofitable. Similarly to the EETT scenario, coal power plants are displaced by gas, renewable and nuclear power plants.

3. Conclusions

The paper tried to answer the question whether the tendency to maintain high share of coal generating in Poland is contradictory to general trends and whether it would contribute relevantly to climate change. The answer for the latter is negative, Polish share in emissions is less than 0.9% thus irrelevant in global scale.

The trend in GHG emissions in Poland is changing, there was a decrease between 1990 and 2000, then the level of emissions was variable but did not present decreasing trend. Only recent five years show decrease, but it is too short to derive any relevant conclusions. In contrary EU28 presents decreasing trend especially in last ten years.

The Polish power generation mix once monopolised by coal, both hard and lignite is gradually changing. Coal will be dominating till 2030 when it is expected to be replaced by nuclear, gas and wind generation.

The quality of air in Poland needs improvement thus low or zero emissions technologies should be applied, also in energy sector.

Acknowledgments
The paper is an effect of the statutory project no. 11.11.210.375

References

[1] IEA Poland 2016 – Energy Policies of IEA Countries – Poland 2016 Review, OECD/IEA, 2017
[2] IPCC 2013, Climate Change 2013 The Physical Science Basis, http://www.climatechange2013.org/images/report/WG1AR5_ALL_FINAL.pdf
[3] IPCC 2014: https://www.ipcc.ch/pdf/assessment-report/ar5/wg3/ipcc_wg3_ar5_summary-for-policymakers.pdf
[4] http://cdiac.ornl.gov/trends/emis/top2014.tot
[5] World Energy Outlook, 2016, IEA Paris
[6] http://ec.europa.eu/eurostat/data/database
[7] Capros P, Paroussos L, Fragkos P, Tsani S, Boitier B, Wagner F, Busch S, Resch G, Blesl M and Bollen J 2014 Description of models and scenarios used to assess european decarbonisation pathways *Energy Strategy Reviews* **2** pp 220–30
[8] Wyrwa A, Pluta M, Skoneczny S, Mirowski T 2014 Modelling the long-term development of an energy system with the use of a technology explicit partial equilibrium model *Vol 8500 of Lecture notes in computer science*, Springer International Publishing pp 489–503
[9] Wyrwa A, Szurlej A, Gawlik L and Suwała W 2015 Energy scenarios for Poland - a comparison of PRIMES and TIMES-PL modeling results *Journal of Power Technologies (Polish Energy Mix)* **95** pp 100-6
[10] Gawlik L (ed) 2013 Coal for Polish energy sector in 2050 perspective - scenario analyses. Mining Chamber of Industry and Commerce (Katowice: Wydawnictwo Instytutu Gospodarki Surowcami Mineralnymi i Energi PAN)

IOP Conf. Series: Materials Science and Engineering **268** (2017) 012004 doi:10.1088/1757-899X/268/1/012004

Chances of coal in European power industry

Zbigniew Łukaszczyk[1] and Henryk Badura[2]

[1] Faculty of Organization and Management, Silesian University of Technology, Roosevelta 26-28, 41-800 Zabrze, Poland
[2] Faculty of Mining and Geology, Silesian University of Technology, Akademicka 2, 44-100 Gliwice, Poland

henryk.badura@polsl.pl

Abstract: Poland's accession to the European Union has reduced the remnants of import barriers. Moreover, the consolidation and commercialization of the energy sector, the implementation of climate package elements and a whole host of other determinants have caused hard coal mining to begin functioning in a highly competitive market, and its negotiating position, as well as the possibility of survival, depends not only on the level of coal prices in international markets, but also on internal competition.

This paper discusses the position of power coal on international markets and presents some current problems concerning the functioning of particular segments of the hard coal market in the European Union and Poland in terms of opportunities and threats that are a result of climate and energy policy.

1. Coal in the world

Coal (hard and brown) is still the most widely used energy source (figure 1). According to BP data [1, 2] its recoverable resources exceed 826.6 billion tons. The advantage of coal, unlike oil and gas, is the presence of its deposits in more than 70 countries around the world, and a widespread access to this feedstock does not create social and political tensions. Only about 17% of this feedstock is the subject of international trade. The rest is used for the production of electric power and heat on the spot – in countries that exploit this coal.

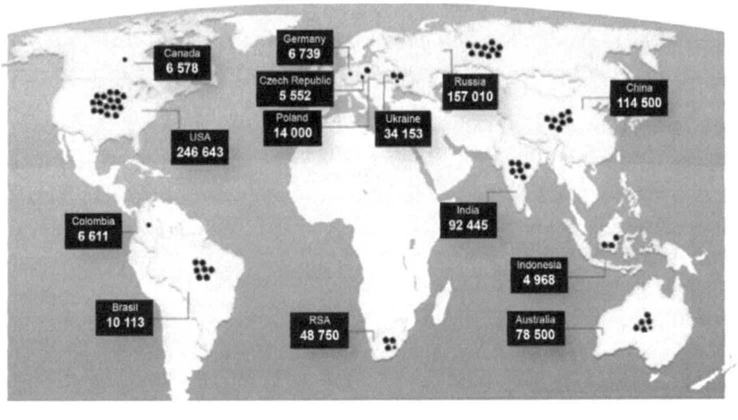

Figure 1. World hard coal reserves (in tons) [1].

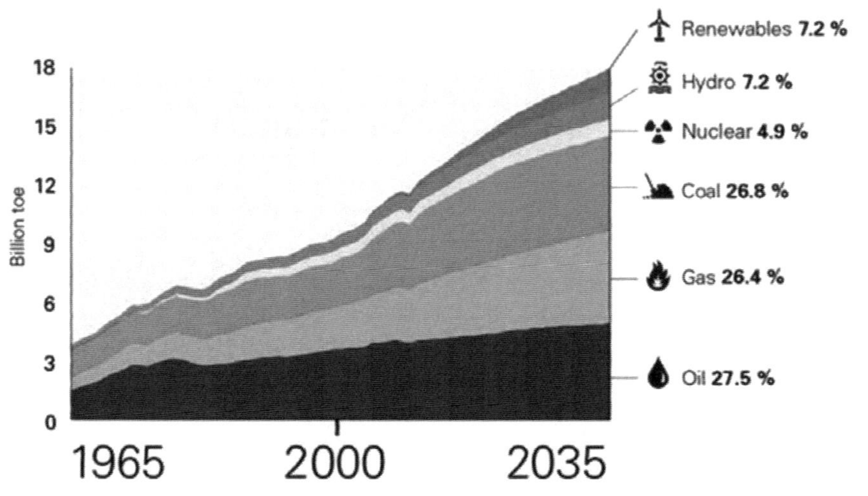

Figure 2. World consumption and forecast of global primary energy demand by source according to BP [1].

Coal is a safe feedstock for transport and storage, and most of all it is cheap and allows generating electricity and heat at a cost acceptable to individual users and the industry, which thanks to it is competitive [3, 4, 5].

On the other hand, the major disadvantage attributed to this feedstock is carbon dioxide emission- because in the context of today's high-efficiency air protection technologies – this greenhouse gas and water vapour are emitted in the combustion process.

In 2014, world hard coal production exceeded 7.1 billion tons and nearly doubled over the past 20 years. The year 2015 is the first year of coal production decrease – 6.9 billion tons (including 5.8 billion tons of power coal – table 1) – caused by the global recession, which led to a decrease in coal production in China and Indonesia.

The downward trend in coal production continued in 2016, when it fell by 458 million tons, which is the biggest decrease since the highest level recorded in 1971. This decrease was mainly the effect of reducing the production in China due to the process of closing down unprofitable and dangerous production units (-320.7 million tons) and in the USA (-141.9 million tons).

Table 1. Production and consumption of power coal in 2015.

2015 [mln Mg]	World	Europe*	Poland*
production	5811	77	59
consumption	5828	228	58
import	1043	176	5.6
export	1003	26	6.9
[percentage]			
production	99.7%	33.8%	101.7%
consumption	100.0%	100.0%	100.0%
import	17.9%	77.2%	9.7%
exports	17.2%	11.4%	11.9%

*power coal

Over the last 10 years, coal trade turnover has increased by about 68% – to about 1 260 million tons (including nearly 300 million tons of coking coal), of which approximately 1 billion tons of coal have been transported by sea. The main contributors to the production, trade and consumption of coal are the Pacific Region countries (non-OECD countries – over 4 950 million tons) [6, 7, 8].

In 2016, the trade turnover of all types of coal increased by 1.9% from the level of 1 308 in 2015 to 1 333.5 million tons, as a result of an increase in power coal exports by 14.6 million tons (i.e. by 1.5%) and the increase in coking coal imports by 10,2 million tons (i.e. 3.4%). At the same time it increased by 21.7% compared to 2010, while exports since the year 2000 increased almost twice, i.e. by 105.3%.

Nowadays more than half of world power coal production comes from China. Since 2009 China has also been the largest net importer of power coal (255.6 million tons /year in 2016), ahead of India (200.1 million tons/year in 2016) and Japan (about 189 million tons/year in 2016).

Table 2. Main producers, exporters and importers of hard coal in 2016 [6].

Main hard coal producers in the world in 2016 (million ton)		Main coal importers in the world in 2016 (million tons)		Main coal exporters in the world in 2016 (million tons)	
China	3242.5	China	255.6	Australia	389.3
India	707.6	India	200.1	Indonesia	369.9
USA	671.8	Japan	189.3	Russia	171.1
Australia	505.3	Korea	134.5	Columbia	83.3
Indonesia	460.5				
Russia	365.5				
RSA	256.9				
Germany	175.6				
Poland	130.9				

2. Primary energy structure in the European Union and Poland

The structure of primary energy sources in countries of the European Union varies considerably as opposed to Poland, where almost 80% of fuels are solid ones (Figures 3, 4, 5).

Energy source	Percentage
Crude oil	9%
Natural gas	14%
Renevable energy	27%
Nuclear energy	29%
Solid fuels	19%
Others	2%

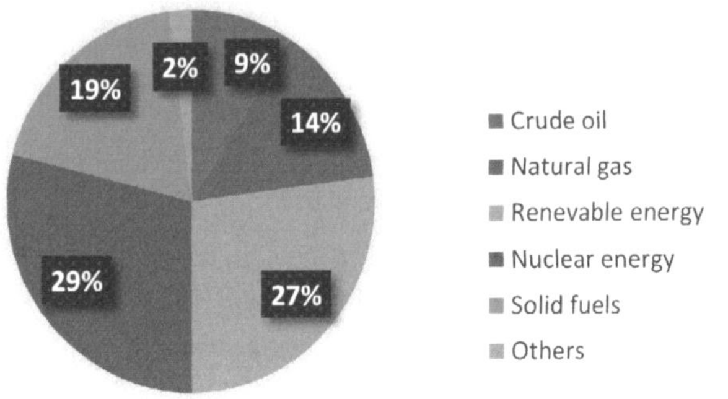

Figure 3. Structure of primary energy sources EU 28 2017.

IOP Conf. Series: Materials Science and Engineering **268** (2017) 012004 doi:10.1088/1757-899X/268/1/012004

Energy source	million toe
Crude oil	613.3
Gas	385.9
Coal	238.4
Nuclear energy	190.0
Hydro energy	78.7
RES	135.6
Total	1641.9

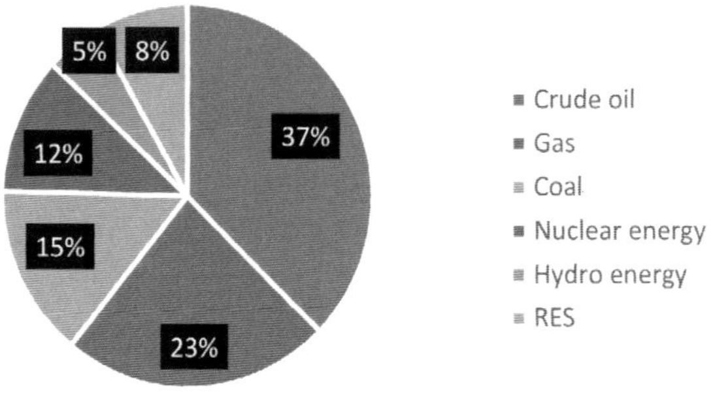

Figure 4. Primary energy consumption in the EU 28 in 2016 [BP 2017].

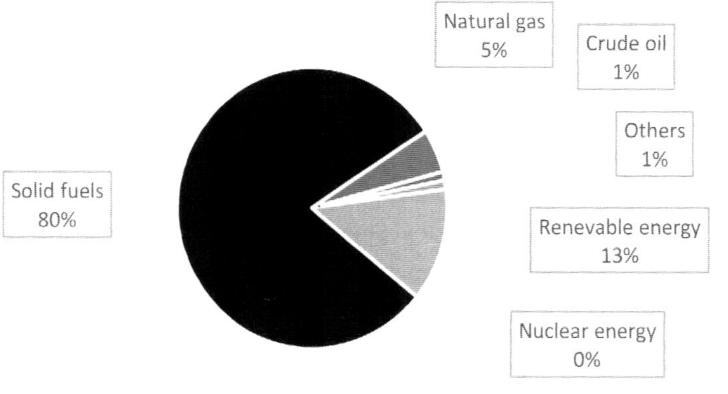

Figure 5. Structure of primary energy sources Poland 2015.

In the European Union primary energy production is dominated by nuclear and renewable energy. The aggregate volume of primary energy production in the EU and Poland is presented in table 3.

Table 3. Primary energy production volume in the EU, in millions of tons of oil equivalent [*Source*: EUROSTAT].

	Aggregate production of primary energy 2003	2015	Nuclear energy	Solid fuels	Natural gas	Crude oil	Renewable energy	Others
EU-28	933.8	765.5	221.02	144.9	111.1	69.1	204.0	15.1
Poland	78.8	67.3	0.0	53.6	3.7	0.9	8.6	0.5

The development of world economy, including the European one, is closely linked to the level of energy consumption. However, there is no doubt that countries having their own primary sources of energy will guarantee themselves a high energy security level and high level of economic independence. On the other hand, EU countries (including Poland) are becoming increasingly dependent on primary energy imports with a distinct decrease in their own production. The dependence on energy feedstocks imports across the EU is around 53%, while Poland with less than 30% dependence situates itself in the top three least dependent European countries – table 4.

Table 4. Primary energy volume in the EU and Poland in millions of tons of oil equivalent [*Source*: EUROSTAT].

Years	2003	2005	2007	2009	2011	2013	2014	2015
			Total imports of primary energy					
EU-28	1350	1463	1467	1495	1433	1441	1413	1477
Poland	32.0	35.6	41.7	42.8	49.0	45.3	46.7	49.9
						Total exports of primary energy		
EU-28	446.7	482.5	483.7	469.8	490.2	533.4	531.0	576.0
Poland	19.9	19.7	17.0	12.8	15.3	20.1	19.6	20.9
						Total production of primary energy		
EU-28	937.0	903.1	861.9	821.6	802.1	790.4	773.0	765.0
Poland	78.8	77.9	71.7	66.8	67.7	70.6	66.9	67.4
						Total consumption of primary energy		
EU-28	1840.3	1883.6	1845.2	1847.8	1744.9	1698.0	1607.0	1626.0
Poland	90.9	93.8	96.4	96.8	101.4	95.8	94.3	95.4

However, for quite some time now, as probably never before on such a scale, we are witnessing a hot and exciting European debate on 'how much trouble we have with all this coal'. Energy feedstocks are and will be a tool of geopolitics and they have their specificity. The energy policy of the European Union (Energy Roadmap 2050) assumes the complete elimination of fossil fuels – not only coal but also crude oil and natural gas – by 2050. Everything is to be clean and safe – but unfortunately it is not clear yet how to achieve this goal without providing a guarantee of energy security for particular countries. Hope lies in future, clean coal technologies. The position of coal in the European Union is being heavily reduced by the implementation of climate policy, the most recent goals of which have been defined at the Paris Conference on Climate Change. The participants of this conference have set up a very ambitious and, according to them, the only effective goal: to maintain global warming well below 2° C in relation to the average temperature of the pre-industrial period. The subject of discussion were the possible steps of actions, including (EU proposal) – a 40% reduction of CO_2 emission by 2030 (with respect to 1990) and the implementation of "zero emission economy" between 2030 and 2050 (which ultimately means the elimination of all fossil fuels, i.e. carbon, natural gas and petroleum products from economy). However, it is important to underline that not all experts agree in the matter of CO_2 emission impact on global warming. These experts believe that human activity has a minor influence on climate changes, which mainly depend on the activity of the Sun. This is evidenced by periods of warming and cooling in the history of Earth, including glacial periods [9].

Over the last 20 years, a total hard coal output in Europe has decreased by more than 25% – in 2015, hard coal production was 98.6 million tons and brown coal production was less than 400 million tons [10].

In the period from 2012 to 2016, the level of coal imports to the European Union decreased by 22%, from 213.5 million tons to 166.8 million tons. Dynamics of imports in individual EU Member States is shown in the table below (table 5).

Table 5. Imports of hard coal to the EU market in 2012-2016 [9].

Years	2012	2013	2014	2015	2016
Million tons	213.5	201.6	204.9	190.7	166.8
Price of power coal USD/t	92.50	94.33	83.93	68.99	59.87

Average price of coal for energy purposes – free-at-frontier of the EU Countries

IOP Conf. Series: Materials Science and Engineering **268** (2017) 012004 doi:10.1088/1757-899X/268/1/012004

The prices quoted in table 5 are the average prices of coal being sold – free-on-rail/port/ land border of the country – importer in the EU. Since 2012 prices have fallen by around 47%. Despite the still high position of coal in the world, its participation in the EU (including Poland) energy policy will depend on the effectiveness of implementation of main climate policy objectives, aimed at reducing the competitiveness of coal by charging it with high costs (including, among others, provision for CO_2 emission prices, subsidizing renewable energy, CCS technology costs, etc.).

3. Polish power coal market

Polish energy sector is based primarily on coal. Its large reserves are located in Poland and account for about 2.2% of global coal reserves. Poland is the largest producer of hard coal in the EU. Hard coal production in Poland accounts for 80% of production in the whole EU.

Poland has a unique structure of energy production in terms of sources, because more than 81% of electric power is produced on the basis of native feedstock, i.e. coal (hard and brown altogether). Such a structure ensures a high level of Poland's energy independence.

In 2016 the utility hard coal-fired power plants produced 81 348 GWh, which means that their production was lower by 0.65% compared to 2015. At the same time, the utility brown coal-fired power plants generated 51 204 GWh last year, which meant that their production was by 4.41% lower than in 2015. In 2016 wind power plants generated 11 623 GWh of energy, i.e. by 15.76% more than a year earlier. Utility gas sources produced 5 776 GWh last year, i.e. 37.77% more than a year earlier. In the whole 2016 (according to data from the Polish transmission system operator) on the functioning of National Power System) electric power production in Poland amounted to 162 626 GWh, which means that it was by 0.53% higher than in 2015. Electric power consumption in Poland in 2016 amounted to 164 625 GWh, and therefore it was by 1.97% higher than in 2015, and the highest since 1988. In 2016 the balance of electric power foreign exchange amounted to plus 1 999 GWh, which means that energy imports were higher than exports.

In the case of Poland, there is no doubt that it is hard coal that will have to be a significant fuel in the energy mix until 2050, even in conditions of low-emission economy forced by the EU. Poland's economy is heavily dependent on coal. We annually import almost 100% of crude oil and petroleum derivatives, about 70% of natural gas (about 10 billion m^3/year) and up to 10 million tons of coal. At current low fuel prices, we spend about 24% of our total revenues from the sum of VAT, CIT and PIT taxes (in the case of fuel prices from the beginning of 2014 it would be over 38%) – and despite that, thanks to coal (Eurostat) Poland occupies a very high 3rd position in the ranking of energy independence. Theoretically we are able to secure 100% of our needs in the area of electricity and network heat production.

Today Poland consumes about 4482.2 PJ (depending on the demand – economic and climatic situation) of hard coal and about 1843.5 tons of brown coal (cumulatively it is an equivalent of about 58 billion m^3 of natural gas). According to PSE reports, more than 60% of electric power is produced from hard coal (34-43 million tons/ year) and more than 17% from brown coal [12]. In addition, about 6% of hard coal-based electric power is produced by industrial power plants. The share of hard coal in the structure of fuels used for the production of network heat is over 70%. For climatic reasons Poland, alongside Denmark, Sweden and Germany, has one of the most extensive heating networks in Europe [13].

Estimation of the coal market in Poland is based on many sources of information [13, 14, 15,16, 17, 18, 19, 20, 21, 22]. Over the last 10 years, total power coal sales have decreased by about 18%, while sales of coal to the domestic market have dropped by about 13%. Sales of thick grades have decreased by about 14%. This is mainly due to the gradual decline in output: 102.24 million tons in 2000 and 70.4 million tons in 2016. Inadequate domestic coal supply, especially heating or coking one, does not compensate demand, which generates its imports. According to ARP [19] in 2016 Poland consumed 78.4 million tons of coal, including 60 million tons of power coal. On the other hand the mining sector sold on the domestic market 51.64 million tons, with imports level of approximately 7.9 million tons (10.3 million tons in total) [13].

IOP Conf. Series: Materials Science and Engineering **268** (2017) 012004 doi:10.1088/1757-899X/268/1/012004

Table 6. Primary energy consumption in Poland [11].

Specification		2007	2008	2009	2010	2011	2012	2013	2014	2015	2016
Primary energy		4119.9	4168.1	3941.2	4338.3	4427.9	4387.1	4429.4	4249.7	4322.9	4482.2
including	Hard coal	2015.6	1961.0	1783.0	2007.9	1955.6	1788.0	1861.8	1725.7	1697.8	1843.5
	Brown coal	499.6	532.7	509.8	484.7	524.4	532.1	549.9	520.3	514.1	490.1
	Crude oil	851.0	894.2	861.8	970.6	1027.2	1068.9	1028.8	1025.6	1109.6	1095.5
	Methane-rich natural gas	429.2	440.5	418.5	448.0	444.5	475.4	479.0	466.6	482.4	520.6
	High-nitrogen natural gas	89.0	84.8	86.5	88.1	93.0	97.3	96.1	94.7	92.2	90.4
	Water. wind. geothermal and solar energy	10.8	11.3	13.3	17.4	20.9	25.6	31.8	37.1	47.3	56.2
	Firewood	135.7	152.6	167.2	180.3	193.1	201.5	211.5	201.4	201.4	204.5
	Refuse derived fuels and other feedstocks	87.0	91.0	101.1	141.3	169.2	198.3	170.5	178.3	178.1	181.4

Since 2013 problems on the domestic coal market have been deepening. It is linked with the gradual decrease in prices of fossil fuels – including coal – on international markets, internal problems and the reduction of electricity production on hard coal as well as the promotion of subsidized renewable energy.

Over 70% of imported power coal is heating coal, distributed through a network of fuel depots in the communal-living sector and other small consumers [19].

3.1. Thermal power industry – electric power production

Thermal power industry, including the production of electricity on the basis of hard coal, brown coal and natural gas, consists of:

- 116 utility power plants and CHP plants, which constitute the national electric power and network heat production base,
- 76 industrial power plants producing about 5% of total electric power,
- Independent CHP plants, producing about 1.5% of total electric power.

Data on the consumption of hard coal in the utility power industry, published among others by the ARP SA (coal sold by the mining sector to the utility power industry) [13] and ARE SA (coal consumed by the utility power industry) [14, 15, 18] are quite divergent as a result of stock levels in power plants and coal imports, which on average accounts for around 3-4% of total hard coal consumption in this sector [19].

IOP Conf. Series: Materials Science and Engineering **268** (2017) 012004 doi:10.1088/1757-899X/268/1/012004

Table 7. Electric power production TWh/year [12].

Years	2007	2008	2009	2010	2011	2012	2013	2014	2015	2016
Total	15.52	155.57	150.91	156.34	163.15	159.85	162.50	156.56	161.77	162.62
Hard coal	93.13	86.55	84.27	89.21	90.81	84.49	84.57	80.82	81.88	81.34
Brown coal	51.14	53.80	50.79	49.60	53.62	55.59	56.96	54,21	53.56	51.20

Depending on economic and climatic conditions, the utility power industry consumes between 34 and 44 million tons/ year of hard power coal fines. Coal stocks kept in piles: from 3.5 to 7 mln tons [13, 14, 15].

By 2020, the consumption in this sector is projected to decrease to 30 million tons/year (increase in brown coal share, subsidized renewable energy, improved efficiency of production, imports of electric power, coal imports, prices of CO_2 emission allowances – EU ETS, IED directives, etc.).

The quality of combusted coal and fuel policy is and will be dependent on the schedule of climate policy implementation:

- mainly coal fine with a low sulfur content <0.6% and a calorific value of 21-24 MJ/kg,
- after 2020, full flue gas desulfurization or high-calorie coal fine (> 27 MJ/kg) with a sulfur content of < 0.4%.

It should be emphasized that coal usually adopted as a benchmark in international trade is power hard coal with a calorific value of 6000 kcal/kg (25.121 MJ/kg), sulfur content of 1.0% and grain size of 0-30 (50) mm [23].

In recent years the situation in hard coal mining sector and especially the level of transaction prices in supplies of coal to the power industry have been strongly influenced by the level of electric power prices in BASE futures, which is being set below the costs of electric power generation on hard coal (subsidized renewable energy, both domestic and imported, mainly from Germany and cheaper energy from brown coal is 150-170 PLN/MWh). This results in exerting pressure of the energy sector (with a price fixing nature) on a significant reduction in hard coal prices – up to 30-40% below import parity level and production costs, including signing long-term contracts which guarantee the power industry lowest purchase prices of coal. In the absence of a cheaper coal purchase option, the utility power industry reduces the generation of electric power on hard coal which in turn results in reduced receptions of this feedstock and thus an increase of the piles volume in coal mines, while maintaining stocks in power plants. Such a situation does not allow for pursuing any reasonable pricing policy by the mining sector.

3.2. Heat production

Heat for heating purposes and technological heat is produced by more than 10 688 heat generation sources, 466 of which are licensed heating companies [14, 18].

According to ARE SA this classification includes:

- 145 CHP plants and utility CHP plants,
- 51 industrial CHP plants,
- 10 120 non-utility CHP plants (small and medium sources),
- 372 manufacturing and distribution companies.

According to ARE SA publication [14] the heating sector annually consumes about 14.5 – 17.5 million tons of hard coal. The stock kept on piles is about 2-3 million tons.

According to forecasts of the Polish Chamber of Commerce, in the years 2020-2030 [24] the demand for coal is expected to decrease by up to 7million tons/year, to 10 million tons/year in the proecological scenario, because of:

- reduction of transmission losses – min. 0.5 million tons/year,
- green heat (RES) – up to 10 million tons/year,
- thermal insulation of buildings – up to 4.5 million tons/year,
- waste incineration – up to 1,0 million tons/year,
- cogeneration.

In 2014, domestic heating companies consumed approximately 17.75 million tons of hard coal. The share of coal in the structure of fuels used for heat production is over 75%. These are mainly coals of low sulfur (< 0.6%) and low ash content.

After 2020, heat generation sources have to introduce full flue gas desulfurization (the alternative may be a high-calorie coal fine > 27MJ/kg with sulfur content < 0.4% – IED Directive).

Forecasts provide a significant increase in the share of renewable energy as well as reduction in transmission losses, thermal insulation of heated buildings and increase of heat generation efficiency.

3.3. Small consumer sector – housing, local boiler houses and heat plants, small industries and farms
According to the Central Statistical Office (GUS), this sector consumes around 11.5-12.5 million tons/year (without network heat) [16, 17, 22]. These are mainly thick and medium grade coals, including about 1.25 million tons of qualified fuels ('eco-pea coals'). This sector is mainly supplied by a network of fuel stores belonging to Authorized Coal Dealers.

It is forecasted that the consumption of coal in this sector will maintain at the current level or decrease by 3-5 million tons per year – in case of radicalization of local governments policy in the field of air protection.

Coal combusted in households and small heat sources (< 1MW) is mainly of high-quality thick and medium grades (about 9 million tons/year) with calorific value of 24 to 31 MJ/kg, sulfur content below 0.6 (0.8) % and ash content less than 10%. However this also includes co-incineration of coal with biomass and municipal waste as well as combustion of low-quality coal fines and coal sludges. Combustion of these fuels, along with road transport, is the main cause of air pollution.

It should be emphasized that the effect of the so-called 'anti-smog law' may be – at best – switching to low-emission techniques with efficiency of about 85% (about 25-35% higher) and thus the increase in demand for EKORET and EKOFINS type fuels that meet technological requirements of low-emission coal boilers, and possible forcing-out coarse grades from the market.

Katowice Coal Holding is actively involved in the works of the Low-emission Coal Boiler Manufacturers Platform and the Qualified Solid Fuels, that acts by the Polish Chamber of Ecology. The aim of the Platform is – among other things – the promotion of low emission technologies for combusting solid fuels, including hard coal, which until recently were utilized in Regional Low Emission Reduction Programs (PONE) and which, according to audits carried out by the Marshall Office in Katowice, were the most effective techniques in relation: financial expenditures – ecological effect.

4. Poland's energy policy until 2050

What can replace coal and what will cover our future energy needs, estimated to be equivalent to another 25 million tons/year of hard coal?

According to the project of 'Poland's energy policy till the year 2050' [25] our future energy needs (table 2) – especially in the area of electric power production – will be covered by nuclear and renewable energy (currently it is mainly wind energy). Sources based on natural gas will mainly stabilize the production of energy from renewable sources.

However, the essential energy production till 2050 is to be still based on coal, mainly hard one. This, however, requires replacing of more than 11 500 MW of power from the oldest coal-fired units with power up to 200 and above 200 MW by the year 2030.

Further use of domestic hard coal means a necessity to conclude a major restructuring of the mining sector (deep costs reduction), massive investments to get access to new resources and launch of mining fronts as well as urgent development of energy policy which will not lead to the marginalization of coal-based energy by heavily subsidized renewable energy and will have mechanisms that allow for a rational financial policy, including taxation, in the mining sector during periods of price booms and boosts. Unfortunately, in extreme boost cases, support for the mining sector could reach the level of at least 1.8 – 2.0 billion PLN/year (about 20 thousand PLN per year per job). Taking into account the EU's ban on public aid for coal, the only option is reduction of mining costs (with estimated coal prices in ARA ports – by at least 26-30 PLN/ton of coal). This option could assure maintaining substantial number of jobs and financial support for the market (consumer spending of workers, services, supplies for mines including: energy, machinery, local and central taxes, etc.). It should be emphasized that coal mining costs, estimated at around 300 PLN/t, include over 100 PLN/t of taxes paid by the mining sector to local and central budget. This should be taken into account when one talks about the competitiveness of coal from Russia, Colombia, and Indonesia. For the money spent on possible coal imports (18-22 billion PLN/year) one could create at least 30 000 new jobs. The argument about extremely harmful effects of coal combustion on our health is important, but it gets more and more outdated. There are practically no power units and larger sources of heat generation that do not have effective air protection installations (smaller sources are successively adapting themselves to the IED Directive). The only gases that are emitted in this case are carbon dioxide and water vapour. Ultimately (CCS) it may be only water vapour.

5. Summary

To promote coal in the EU today one needs a lot of courage, but also a lot of imagination to answer the question WHAT INSTEAD OF COAL?

From the perspective of Polish economy, including energy security, the most rational seems to be the coal option based on low-emission technologies with high efficiency of generation (45% and above), energy from renewable sources, including prosumer energy, supported by energy from gas sources and nuclear power industry. The lack of any element of this production structure (i.e. delay in its implementation) in the forecasted period may mean a significant limitation in production and supply of electric power and heat.

References

[1] BP Energy Outlook 2035 2015 **2**.
[2] Word Energy Outlook 2014 IEA.
[3] Kurczabiński L and Łukaszczyk Z 2015 Czysty węgiel – czyste środowisko Nasz Holding **2**
[4] Kurczabiński L and Łukaszczyk Z 2015 Role of the Coal in Polish Energy Sector. Meeting of the Group of Experts, Committee on Soustainable Energy, UNECE, Geneva.
[5] REMAP IRENA 2015 Perspektywy rozwoju energii odnawialnej w Polsce.
[6] Gilecki R 2014 Sektor energii świata i Polski. Początki, rozwój i stan obecny Word Energy Council Polski Komitet Światowej Rady Energetycznej Warszawa.
[7] Korzeniowski M and Kurczabiński L 2014 Analiza problemów krajowego rynku węgla. The paper of the XXVIII Conference in the series of Issues of Energy Resources and Energy in the National Economy Zakopane.
[8] EURACOAL 2015 Market Report **1**, **2**.
[9] Kurczabiński L 2016 Polski węgiel po Konferencji Klimatycznej – Paryż 2015 Ekologia **1**
[10] EURACOAL 2013 Coal in Europe 2013.
[11] Popkiewicz M 2016 Rewolucja energetyczna – ale po co? (Katowice: Sonia Draga Sp. z o.o.)
[12] Polskie Sieci Elektroenergetyczne SA. Miesięczne raporty z RB i KSR. 2006-2016.
[13] Industr. Development Agency JSC 2016 Podstawowe informacje o rynku oraz sektorze węgla kamiennego w Polsce w latach 2012-2015 Katowice.
[14] ARE SA 2015 Statystyka Ciepłownictwa Polskiego 2014 Warszawa.
[15] ARE SA 2015 Statystyka Elektroenergetyki Polskiej 2014. Warszawa.
[16] GUS (Central Statist. Office of Poland) 2015 Zużycie energii w gospodarstwach domowych

IOP Conf. Series: Materials Science and Engineering **268** (2017) 012004 doi:10.1088/1757-899X/268/1/012004

Warszawa
[17] GUS (Central Statist. Office of Poland) Zużycie paliw i nośników energii w 2014 Warszawa
[18] Energy Regulatory Office 2015 Energetyka cieplna w liczbach – 2014 Warszawa
[19] Industr. Development Agency JSC 2016 Import i przywóz węgla (nabycie wewnątrzunijne) na obszar Polski w okresie 2014 i 2015 Katowice
[20] ARE SA 2015 Katalog elektrowni i elektrociepłowni zawodowych – 2014 Warszawa
[21] ARE SA 2015 Katalog elektrociepłowni przemysłowych – 2014 Warszawa
[22] Zamieszkane budynki. Narodowy Spis Powszechny Ludności i Mieszkań 2011 GUS Warszawa 2013
[23] PLATTS Mc Graw Hill Financial ICR Coal Statistics Monthly 2010-2050.
[24] Regulski B 2009 Ciepło systemowe na rynku energii – aspekty węglowe Izba Gospodarcza Ciepłownictwo Polskie Warszawa
[25] Ministerstwo Gospodarki 2015 Polityka energetyczna Polski do roku 2050 Projekt Ministerstwo Gospodarki Warszawa

IOP Conf. Series: Materials Science and Engineering **268** (2017) 012005 doi:10.1088/1757-899X/268/1/012005

The possibility of increasing the efficiency of accessible coal deposits by optimizing dimensions of protective pillars or the scope of exploitation

Piotr Bańka, Henryk Badura and Marek Wesołowski

Faculty of Mining and Geology, Silesian University of Technology, Akademicka 2, 44-100 Gliwice, Poland

piotr.banka@polsl.pl

Abstract. One of the ways to protect objects exposed to the influences of mining exploitation is establishing protective pillars for them. Properly determined pillar provides effective protection of the object for which it was established. Determining correct dimensions of a pillar requires taking into account contradictory requirements. Protection measures against the excessive influences of mining exploitation require designing the largest possible pillars, whereas economic requirements suggest a maximum reduction of the size of resources left in the pillar. This paper presents algorithms and programs developed for determining optimal dimensions of protective pillars for surface objects and shafts. The issue of designing a protective pillar was treated as a nonlinear programming task. The objective function are the resources left in a pillar while nonlinear limitations are the deformation values evoked by the mining exploitation. Resources in the pillar may be weighted e.g. by calorific value or by the inverse of output costs. The possibility of designing pillars of any polygon shape was taken into account. Because of the applied exploitation technologies the rectangular pillar shape should be considered more advantageous than the oval one, though it does not ensure the minimization of resources left in a pillar. In this article there is also presented a different approach to the design of protective pillars, which instead of fixing the pillar boundaries in subsequent seams, the length of longwall panels of the designed mining exploitation is limited in a way that ensures the effective protection of an object while maximizing the extraction ratio of the deposit.

1. Introduction

The essential way of protecting important objects exposed to mining exploitation is designing the protective pillars. Exploitation of a deposit in the protective pillar, basically acceptable, requires special conditions, such as selection of the appropriate method of roof control, an adequate time coordination of exploitation fronts, abandonment of the exploitation of some seams and an appropriate objects' protection. It is also important to develop an accurate forecast of the influences of mining exploitation, based on one of the newest methods [e.g. 1]. It should be noted however that establishing protective pillars is not a flaw-free way of protecting objects. Leaving an unextracted part of seam entails a number of negative effects, both in rock mass and on the terrain surface. These include: the increase of the stresses value and risk of high-energy shocks and bumps in the part of deposit located within the pillar as well as formation of a relative upheaval on the terrain surface.

It is very important to design properly the boundaries of a protective pillar in individual seams. A pillar that is too small not only fails to protect the object but can even worsen its maintenance conditions in comparison with mining out of the deposit occurring under the object. On the other hand, a protective pillar that is too large (in terms of ensuring that the permissible deformation values are not exceeded) causes a deterioration of economic results of a mine due to higher operating costs of hampered mining (inside the pillar) compared to the off-pillar exploitation.

Over the last few decades the method of determining pillars in coal mines has undergone some changes, which were introduced in the published instructions. Older Instructions were put into mandatory application by the regulations of Minister of Mining and Energy (Instructions no. 19 of 1961, Directive no. 4 of 1986).

The latest instruction for determining protective pillars, developed in Central Mining Institute [2], provides two ways of determining protective pillars:

- in a manner similar to the Instruction of 1961, the pillar's boundaries are determined in individual seams based on the angles of range of permissible exploitation influences. The size of this angle depends on the resistance category of an object (or objects).
- with the use of a simplified analytical method that ensures that the object or objects are fully protected from the occurrence of deformations that exceed permissible values. This method consists in the determination of pillar radiuses based on geological and mining data, basic parameters of the theory and specific criterion indicators of deformation that characterize resistance of the protected object.

The guidelines also allowed the determination of a protective pillar on the grounds of an expert opinion based on other justified assumptions. The solutions presented in this article fit in the above, expert way of designing the boundaries of a protective pillar.

2. Determining the boundaries of the protective pillar (the scope of exploitation) as the solution of a certain nonlinear programming task

Designing of the protective pillar (the lengths of longwall panels) was treated as a nonlinear programming task [3]. Solving it demands searching for the minimum of the objective function F_c:

$$F_c = \sum_{i=1}^{n} w_i g_i P_i \tag{1}$$

with limitations:

- for surface objects:

$$\sum_{i=1}^{n} \varepsilon(R_i) \leq \varepsilon_{\text{dop}} \tag{2}$$

$$\sum_{i=1}^{n} T(R_i) \leq T_{\text{dop}} \tag{3}$$

$$\sum_{i=1}^{n} K(R_i) \leq K_{\text{dop}} \tag{4}$$

$$R_{i\,\text{grn}} \leq R_i \leq R_{\text{maks}} \tag{5}$$

- for shafts (pits):

$$max\left|\sum_{i=1}^{n} \varepsilon_{zi}(z, R_i)\right| \leq \varepsilon_{z\,\text{dop}} \tag{6}$$

$$R_{i\,\text{min}} \leq R_i \leq R_{i\,\text{maks}} \tag{7}$$

Additionally, in the case of determining optimal dimensions of protective pillars, it is possible to introduce an additional limitation of the form:

$$R_{i+1} \geq R_i \quad \text{for} \quad i = 1, \dots, \text{n} - 1 \tag{8}$$

In the case of optimization of the lengths of longwall panels this limitation looks as follows:

$$R_i \geq R_j \quad \text{for} \quad i, j \in \langle 1, \dots, n \rangle \tag{9}$$

where:
n – number of seams (or longwalls which panel length is subject to optimization),
w_i – weight of the i-th seam (the i-th longwall),
g_i – thickness of the i-th seam (height of the i-th longwall),

P_i – surface area of the pillar in the i-th seam, (surface area of unextracted part of the i-th longwall panel),
ε_i – maximum horizontal strain caused by exploitation of the i-th seam (the i-th longwall),
T_i – maximum inclination caused by exploitation of the i-th seam (the i-th longwall),
K_i – maximum vertical curvatures caused by exploitation of the i-th seam (the i-th longwall),
ε_{zi} – maximum vertical strain at depth z caused by exploitation of the i-th seam,
ε_{dop}, T_{dop}, K_{dop}, ε_{zdop} – values of permissible deformations,
R_{igrn} – value of the most unfavorable length of radius (side) of the pillar (in the case of lengths of longwalls panels optimization $R_{igrn} = 0$),
R_{imin}, R_{imaks} – lower, upper limit of variation of the pillar radius in the i-th seam (in the case of lengths of longwalls panels optimization R_{imaks} equals to the panel length of the i-th longwall).

Limitation (8) allows designing pillars which radiuses increase regularly with the increase of the depth of seams occurrence. In the case of abandoning this limitation, the optimized protective pillar may have a shape that excludes safe off-pillar exploitation (figure 1). Resignation from the use of limitation (8) requires checking if the exploitation of lower occurring seam does not lead to destruction of a higher occurring one.

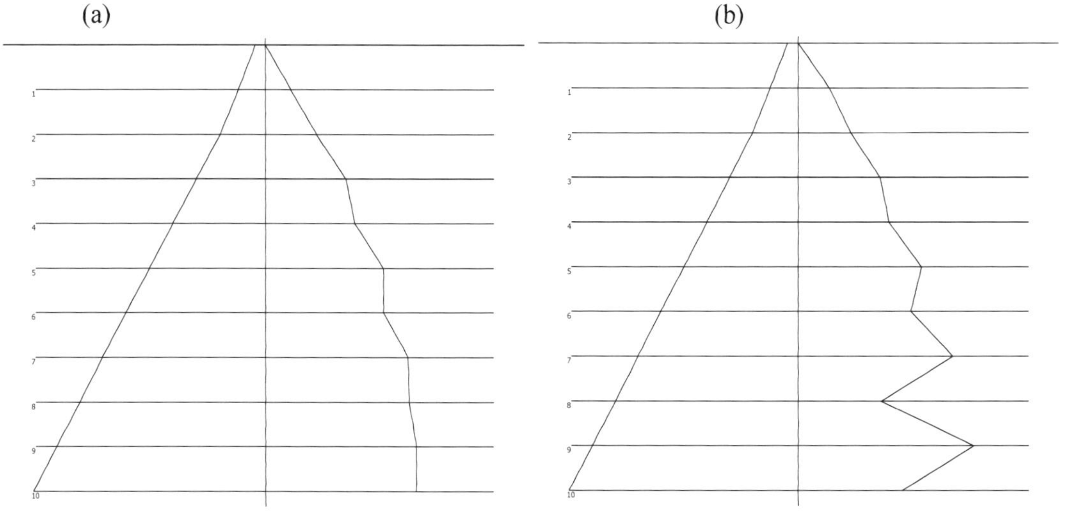

Figure 1. Diagram of radiuses of protective pillar determined in accordance with Regulation No. 19– left side of figures (a) and (b) and optimized – right side of figures (a) and (b); (a) – including limitation (8), (b) – without limitation (8).

Due to the fact that when exploitation approaches to the shaft (pit), the maximum of vertical deformations 'moves' towards the seam plane and it may happen that for a certain section of the shaft, vertical deformations occurring before the exploitation reaches a predetermined distance, will exceed permissible values even though their final values will be lower than permissible ones. Therefore the calculation of vertical strains is performed with the use of the following scheme (figure 2).

For a given radius of the pillar R, the depth of seam occurrence H, the radius of the range of influences r_z, the value of the pillar's radius is determined, hereinafter referred to as R_{kryt}, for which the off-pillar exploitation evokes maximum vertical strains at a given level z. In the case when radius of pillar R is smaller than R_{kryt} determined for the examined level z, vertical strains are calculated with the assumption that $R = R_{kryt}$, whereas in the case when radius of pillar R is larger than R_{kryt}, calculation of vertical strains is performed with the use of the actual value of radius R. The depth z, for which $R = R_{kryt}$, is hereinafter referred to as z_{kryt}. Thin lines in figure 2 represent the course of ε_z (R, z) for R varying from 200 m to 40 m. The dashed horizontal line represents the value $z = z_{kryt}$. For the value $z \leq z_{kryt}$, the actual value of R is taken into account for calculations, while for $z > z_{kryt}$, calculations are performed with the use of R_{kryt} value. The resulting course ε_z (R, z) takes the form of an envelope of curves ε_z (R, z) - marked in figure 2 with a thick line.

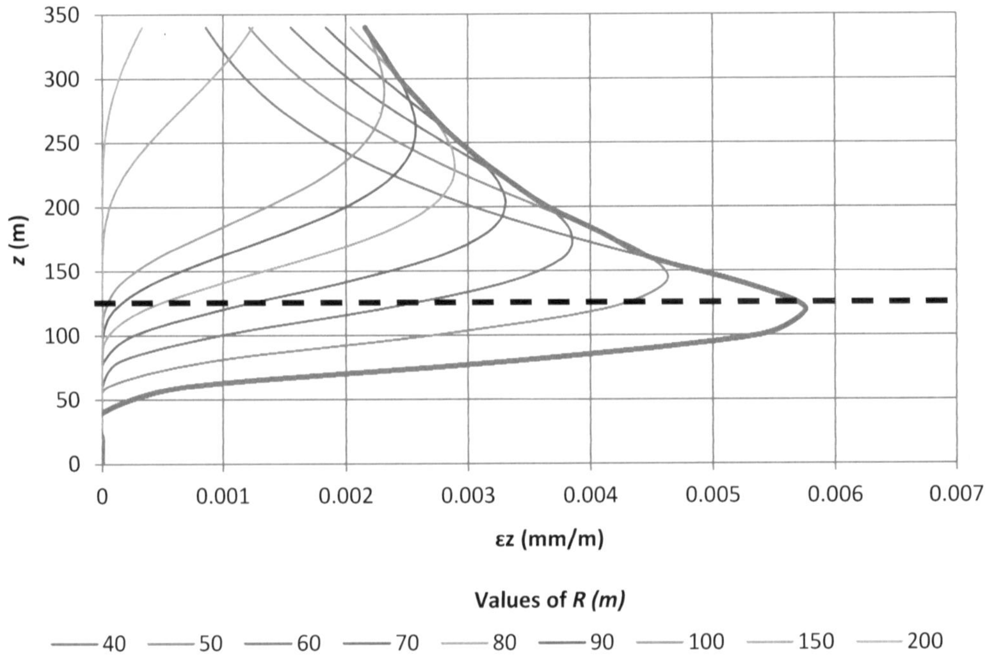

Figure 2. Method adopted for determining ε_z (R, z).

The developed algorithms assume the possibility of extracting certain seams within the boundaries of the protective pillar (seams for which the maximum deformation indicator values for the half-plane exploitation are lower than permissible ones), which reduces the negative effects of pillar placement, such as relative terrain upheavals and an increase of stresses in rock mass. Also a possibility of imposing pillar radius size in selected seams was considered – which allows leaving the binding pillar boundaries in some seams and adjusting pillar size in the other seams.

In the case of determining a protective pillar for a shaft (pit), it is possible to assume different permissible deformation values for different shaft depths, while during the optimization of a pillar for the group of surface objects it is possible to assume various permissible deformations for individual objects.

The developed algorithms and programs make it possible to determine, separately for each seam, the 'resourceless zones', i.e. parts of seams in which exploitation will not be conducted, regardless of the size of protective pillar. Such zones may be for example the areas of geologic disturbances. In particular, a 'resourceless zone' may be constituted of abandoned workings of the accomplished mining exploitation, the influences of which have emerged before the erection of an object for which the protective pillar is determined or were taken into account during the assessment of object's resistance.

The problem of determining the protective pillar can also be approached in a different way than the one presented above. Instead of determining boundaries of the protective pillar in successive seams, the scope of the planned exploitation (the lengths of longwall panels) can be optimized to protect the object against influences of mining exploitation. In this approach, the planned longwalls are divided into two groups: fixed longwalls, which panels length is not subject to change and variable longwalls, which panel lengths are subject to optimization. All plots are divided into rectangles of fixed width. The influences of mining-out the subsequent rectangle are then calculated (i.e. the increase of the values of deformation indicators as well as the extreme values of indicators that have occurred since the beginning of exploitation of the given longwall) for subsequent calculation points. In the case of variable longwalls, the deformation indicator values are calculated for each successive rectangle. While seeking the minimum of an objective function, the limitations (2), (3), (4) and (9) are taken into account, which allows planning the scope of exploitation that does not exclude the subsequent extraction of the remaining deposit.

For seeking the minimum of an objective function (1) the method of a shifting penalty function [3] is used. The above method belongs to the group of approximation methods that solve a sequence of substitutive problems in such modified form, that in subsequent iterations the solutions of these problems are more and more approximated to the solution of the initial problem. Solving of substitutive problems was performed with the use of Powell method, which belongs to the group of non-gradient methods of improvement, i.e. methods determining the solution of a nonlinear programming problem by searching the R_n space in appropriately constructed directions that create the basis of this space. In turn, the minimization in these directions is performed by the non-gradient method of square approximation.

3. An example of determining the protective pillar and optimizing the scope of exploitation

The results of sample calculations of the protective pillar size (the lengths of longwall panels) performed according to the presented methodology are shown below. The simplest situations – determining the optimum protective pillar size for a point object and shaft – were omitted, because of they were already presented in earlier papers [e.g. 4, 5].

3.1. Example 1. Determining the protective pillar for a group of objects with different resistance to the influences of mining exploitation

Figure 3 shows a schematic representation of a surface map showing the objects to be protected. The boundary of protected area was marked too.

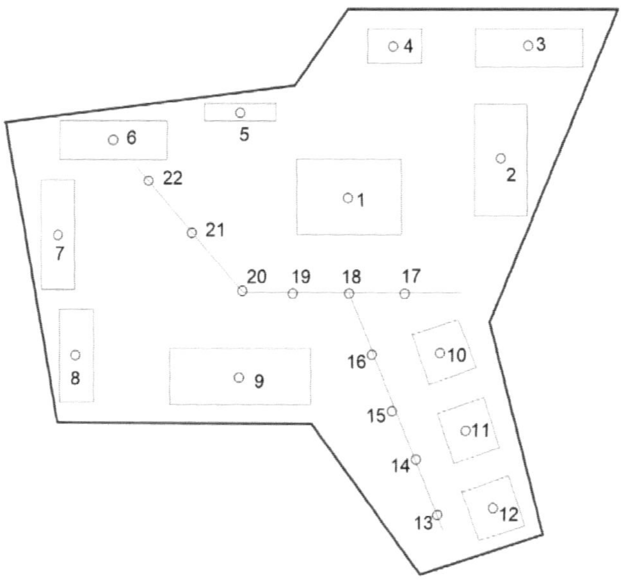

Figure 3. Boundaries of the protected area, location of protected objects.

Permissible values of horizontal deformations and slopes for protected objects are shown in table 1.

Table 1. Permissible values of deformation indicators.

Object number	Permissible horizontal strains (mm/m)		Permissible inclinations (mm/m)	
	ε_x	ε_y	T_x	T_y
1	1.5	1.5	2.5	2.5
2, 3, 5, 6, 7, 8, 9	3	3	5	5
4, 10, 11, 12	6	6	10	10
13, 14, 15, 16, 21, 22	6	3	10	5
17,18, 19, 20	3	6	5	10

It was assumed that under protected objects there occur nine coal seams that would be subject to mining exploitation. Furthermore it was presupposed that the exploitation would be conducted with roof rocks caving and that the resources in each seam are characterised by the same weight. Data on seams is presented in table 2. Table 2 also shows the calculated values of pillars radiuses in individual seams. The values of radiuses determined in accordance with Regulation No. 19 are also included for comparison purposes.

Table 2. Data on seams, results of calculations of pillar radiuses.

Seam number	Depth of occurrence (m)	Seam thickness (m)	Pillar radius (m)	
			According to Regulation No. 19	Optimized pillar
1	370	1.5	231	280
2	470	1.5	294	350
3	540	2.0	338	360
4	680	2.4	425	430
5	750	3.0	469	430
6	800	4.2	500	450
7	900	4.8	563	460
8	930	2.0	581	466
9	960	6.0	600	470

The results of optimization are illustrated in figure 4. The "optimized" pillar is characterized by larger radiuses in seams occurring at shallower depths and smaller ones in the case of seams occurring deeper, compared to the values determined in accordance with Regulation No. 19. This trend is in accordance with the results of observations, which indicate that pillars designed with the use of Regulation No. 19, in the case of shallowly occurring seams, were often turning out to be too small.

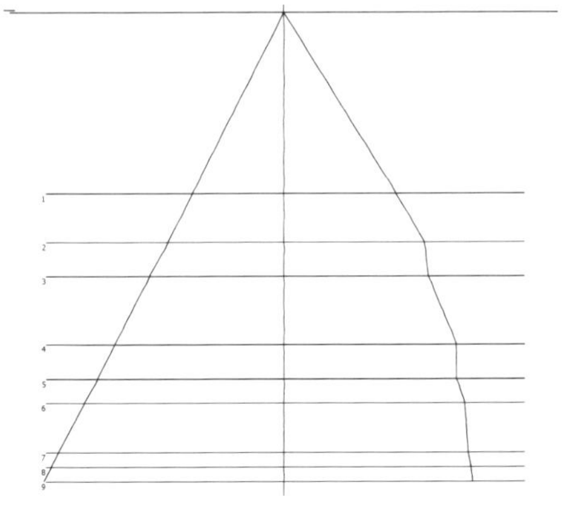

Figure 4. Diagram of radiuses of the protective pillar determined in accordance with Regulation No. 19 (left side) and the optimized one (right side).

3.2. Example 2. Optimizing the scope of exploitation

In the presented test it was assumed that the protected object (marked with the black circle in figures 5 and 6) would be affected by the influences of mining exploitation designed in two seams marked with numbers 1 and 2. Seam 1 occurs at the depth of 400 m whereas seam 2 occurs at the depth of 500 m. The thickness of seams is 2 m and 3 m respectively. The exploitation takes place in the entire thickness of seams. In seam 1 it was planned to extract the deposit with longwalls 1 ÷ 4 (figure 5a). In seam 2 it is planned to mine longwalls 5 ÷ 8 (figure 5b). It was assumed that deformations occurring in the protected object could not be greater than those permissible for the II category of protection.

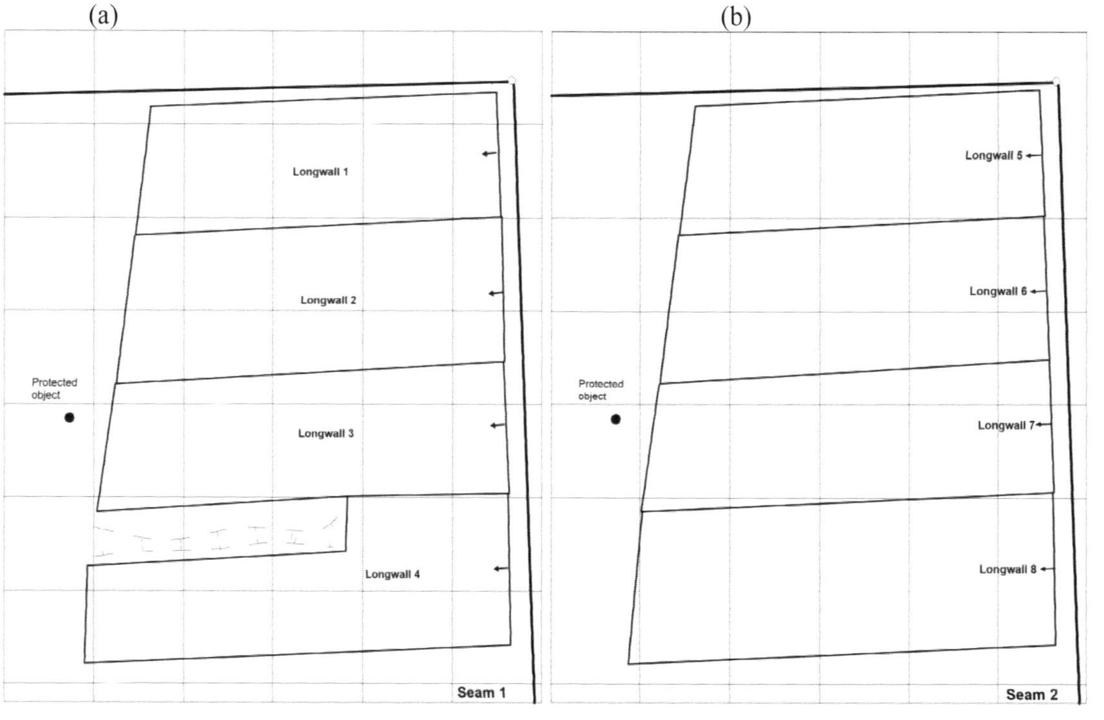

Figure 5. Map of the seam 1 – (a), map of the seam 2 – (b).

The following limitations were imposed on panel lengths of the designed longwalls:

- panel length of longwall 2 should be smaller than panel length of longwall 1,
- panel length of longwall 3 should be smaller than panel lengths of longwalls 2 and 4,
- panel length of longwall 6 should be smaller than panel length of longwall 5,
- panel length of longwall 7 should be smaller than panel length of longwall 6,
- panel length of longwall 8 should be smaller than panel length of longwall 7.

As the result of the conducted optimization, panel lengths of the designed longwalls providing maximization of deposit utilization and at the same time ensuring that the permissible deformations in the area of protected object would not be exceeded were set down. Determined values of panel lengths of the designed longwalls are shown in table 3.

The results of optimization are illustrated in figure 6. The permissible scope of exploitation of the deposit with longwalls 1 ÷ 8 is shown as shaded. Parts of panels of the designed longwalls that should not be extracted to protect the object remained unfilled.

Table 3. Results of calculations of the designed longwalls panel lengths.

Longwall number	Designed panel length (m)	Optimal panel length (m)
1	700	685
2	730	626
3	790	583
4	820	800
5	700	681
6	730	616
7	790	605
8	830	598

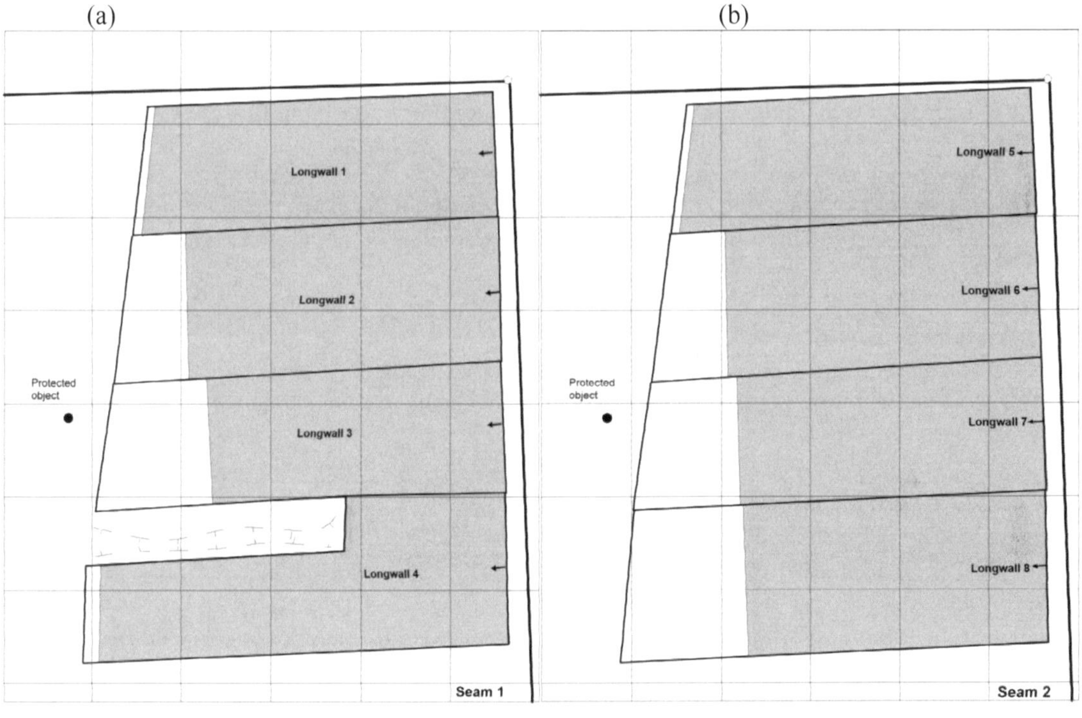

Figure 6. Optimal panel lengths of the designed longwalls in the seam 1 – (a), in the seam 2 – (b).

The adopted limitations for panel lengths of the designed longwalls allowed obtaining a solution that was acceptable due to the requirements connected with the prevention of rockbursts. It is not ruled out that the remaining parts of the deposit will be extracted later, when the protection of an object is not required anymore.

4. Summary

The decision of establishing a protective pillar, due to many negative consequences of this fact, should be preceded by a careful analysis of the need to set it up. There is no doubt that abandoning the exploitation at sufficiently large distance from the object (i.e. setting up a properly designed protective pillar or determining a suitable scope of exploitation) is a way that enables full protection of the object, even in the case of changes in exploitation plans of a mine caused by, for example, mining disasters or disturbances in seam occurrence. In the case of under-object mining (a variant without leaving a protective pillar) stopping of the exploitation front, forced by objective causes, may cause damage to the object.

In the case when a pillar is necessary, it is very important to design it properly, so that it can effectively protect the object(s) for which it was established. Protective pillars determined with the use of this method both effectively protect the object and ensure the fulfilment of the criterion of minimizing the amount of resources left in the pillar. In many cases the equally effective solution as setting up of a protective pillar may be optimization of panel lengths of the designed longwall workings.

References

[1] Gruchlik P, Kowalski A, Rajwa S and Walentek A 2014 Modelling of rockmass and surface deformation caused by roadway mining in light of geodetic observations *15. Geokinematischer Tag, 15. und 16. Mai. Freiberg* pp 107-16

[2] Jędrzejec E, Kowalski A and Kwiatek J 1996 *Determination of protecting pillars for surface buildings, shafts and pits within the boundaries of mining areas of hard coal mines (theoretical foundations and guidelines)* (Katowice: Central Mining Institute)

[3] Findeisen W, Szymanowski J and Wierzbicki A 1980 *Theory and calculation methods of optimization* (Warsaw: Polish Scientific Publishers)

[4] Drzęźla B and Bańka P 1994 Determination of protecting pillar dimensions for surface building and shafts using non-linear method of programming *Work Safety and Environmental Protection in Mining* **4** pp 39–44

[5] Drzęźla B, Białek J and Bańka P 1995 Protecting pillars – circular or square? *Mining Review* **5** pp 1–4

IOP Conf. Series: Materials Science and Engineering **268** (2017) 012006 doi:10.1088/1757-899X/268/1/012006

An overview of potential benefits and limitations of Compressed Air Energy Storage in abandoned coal mines

Marcin Lutyński

Faculty of Mining and Geology, Silesian University of Technology, Akademicka 2, 44-100 Gliwice, Poland

marcin.lutynski@polsl.pl

Abstract. Compressed Air Energy Storage (CAES) is one of the methods that can solve the problems with intermittency and unpredictability of renewable energy sources. The storage is charged by increasing air pressure with the use of electrically driven compressors, which convert the electric energy into potential energy. The pressurized air is stored in compressed air storage volumes (caverns, voids, porous structures etc.) of any kind and can then be released upon demand to generate electricity again by expansion of the air through an air turbine or gas turbine. Limited availability of salt caverns in Europe creates difficulties in the implementation of this concept on larger scale. This paper deals with underground storage part in CAES concept and lists benefits related to the storage of air in abandoned coal mines. Examples of natural gas storage in abandoned coal mines are given and compared with the compressed air storage. The study shows an example of coal mine volume calculation. The non-exhaustive list of problems and solutions associated with this idea is given in order to develop this concept at larger scale.

1. Introduction

An increased share of renewable energy, particularly wind power and photovoltaics, in an energy mix of European countries create difficulties for the regional stability of the electricity grid on the supply side. The reason of this is the constant growth of the share of wind power and photovoltaics in the last ten years. In total, between 2004 and 2015, the share of renewable energy almost doubled, reaching 16.7% of gross final energy consumption in 2015, at the same time wind and solar energy are growing the fastest. In 2015, the EU generated 26.0 million tons of oil equivalent (Mtoe) from wind energy — a more than five-fold increase compared with 2004. In the same year, solar energy contributed 13.1 Mtoe which is more than 18 times as much as in 2004, see figure 1 [1]. Wind and solar generation both experience: intermittency; a combination of regional meteorological uncontrolled variability and local unpredictability, and dependency on resources that are location dependent. Intermittent renewables are challenging because they disrupt the conventional methods for planning the daily operation of the electric grid. Power from these sources fluctuates over multiple time horizons, forcing the grid operator to adjust its day-ahead, hour-ahead, and real-time operating procedures. On the demand side, sudden increase of electricity need cannot be covered by renewables and requires a permanent stand-by of conventional power suppliers. These vital supply and demand aspects along with distribution inter-regional aspects are posing serious challenges for generation authorities and grid operators during the course of integrating wind and solar generation with the conventional grid [2].

IOP Conf. Series: Materials Science and Engineering **268** (2017) 012006 doi:10.1088/1757-899X/268/1/012006

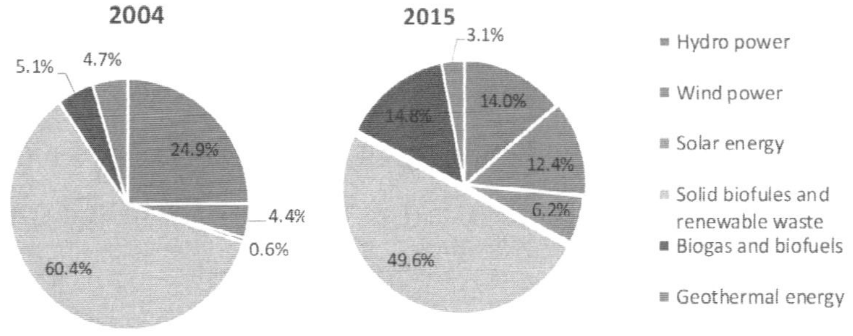

Figure 1. Gross inland consumption of renewable energy, by source, EU-28, 2004 and 2015 [1].

The most important aspect of this variation in power generation is the meteorological variability in power supply. Even if operators could predict the output perfectly, it still poses specific challenges to the grid operator. As a consequence on regional scale, the seconds to minutes time scale, grid operators must deal with fluctuations in frequency and voltage on the transmission system that, if left unchecked, would lead to blackouts. Therefore, operators supply active or reactive power into the grid for a supply/demand balance based on forecasted power generation. This is necessary to maintain a stable and safe frequency and voltage profile in the grid. In addition, with Photovoltaic and wind, two situations are possible when the maintenance of reserves that stand ready to provide additional power. When renewable energy generation produces less energy than predicted, one needs to ensure the availability of a dispatchable load to be 'exploited'. When excess power of renewable energy generation produces more energy than the predicted energy demand, the surplus energy needs to be stored. This is particularly challenging in a system where an inelastic demand has to meet supply in real-time, in the presence of only limited storage capacity.

This issue is referred to as the 'incompressibility of power systems' and is observed in different European electricity markets such as Belgium, Germany, France and the Netherlands, with hours showing negative electricity prices, as well on day-ahead, intra-day and real-time balancing markets. These negative prices show inflexibility of the grids, and therefore the difficulty of the system to cope with periods of excess generation. This is particularly difficult and seems irrational from the economical and engineering point of view when the energy producer has to pay for the electricity supplied [3].

The amount of wind and solar resources are on preferred locations. Unlike coal, gas, oil or uranium, it cannot be transported to a generation site that is grid-optimal. Generation must be co-located with the resource itself, which are often far from the load centers where the power is to be eventually used. New regional transmission capacity is often required to connect wind and solar resources to the rest of the grid. Transmission costs are especially important for offshore wind resources, and such lines often necessitate the use of special technologies not found in land-based transmission lines. Integrating wind and solar generation resources into the conventional electricity grid involves managing other controllable operations that may affect many other parts of the grid, including conventional generation [4].

Above-mentioned problems associated with renewable energy source and power instability will have a dramatic impact on future power generation, in particular for a) the countries where coal is a dominant energy source and b) the European areas with a less predictable regional weather forecast. Recent key findings of the report prepared by 32 companies and supported by EU shows that 'both power-to-power storage and conversion to other carriers have the potential to play an important role in providing flexibility to the power system. They will make it possible to ensure that large amounts of renewable energy are not wasted, but are used to reduce the amount of required non-renewable energy sources generation and decarbonised heating, transportation and gas grid expansion' [5]. It is anticipated that demand for power storage will grow up to 10 times until 2050. In line with the EU policy to reduce fossil fuel energy and secure the supply against international disagreements (i.e. Russian gas deliveries), existing energy infra-structure at regional level must be secured, improved and expanded.

One of the viable options is CAES – Compressed Air Energy Storage, a technology where vast amounts of air can be compressed and stored under pressure in existing underground caverns and mines. When electricity is required, the pressurized air is released, heated and expanded in an expansion turbine driving a generator for electricity production. This paper reviews current technologies of CAES and provides an overview of potential benefits and limitations of compressed air storage in abandoned coal mines.

2. Basic concept of CAES

The concept of large-scale compressed air storage was developed in the middle of the last century. The first patent on CAES in man-made cavities, as a means of storing electrical energy was issued in 1948 in US by Frazer W. Gay [6]. In the manufacturing industry, compressed air is broadly used either as an energy carrier for pneumatic processes (such as drilling and/or carving) or it serves as a process fluid carrier (e.g. for cleaning or varnishing). Either way, compressed air is generated almost exclusively on site by employing electrical energy. For example, only in Germany, currently 16 TWh of energy are consumed annually to provide compressed air for industrial purposes, which amounts to about 2.5% of the German overall electricity consumption [7].

The overall concept of CAES, which is used in nowadays industrial surface facilities is well known. The storage is charged by increasing air pressure with the use of electrically driven compressors, which convert the electric energy into potential energy (exergy). The pressurized air is stored in compressed air storage volumes (caverns, voids, porous structures etc.) of any kind and can then be released upon demand to generate electricity again by expansion of the air through an air turbine or gas turbine (figure 2).

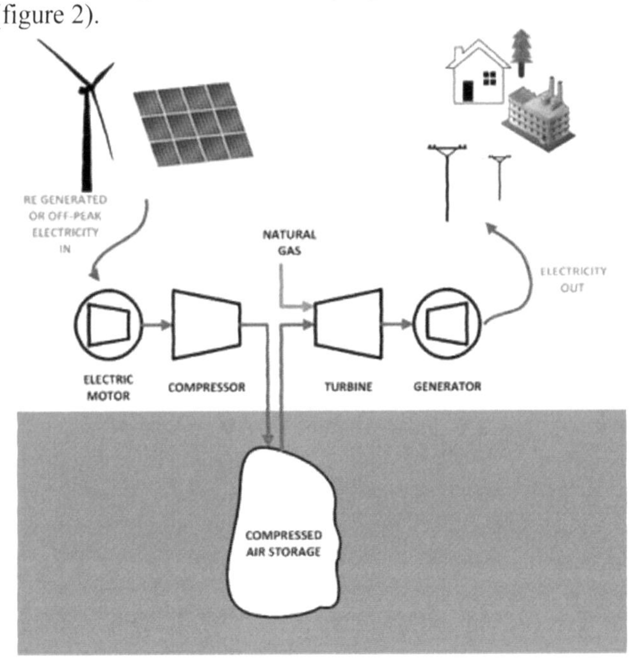

Figure 2. General concept of CAES.

There are several concepts existing today at different levels of development. In general, the concepts can be divided into diabatic, adiabatic and isothermal CAES. The different options are at three stages of technological readiness levels. The diabatic systems are proven, can be up-scaled and can be rolled out for salt domes. The adiabatic and isothermal systems however, are proven as a technology and need to be up-scaled and revised to be implemented in existing large space sub-surface infrastructures.

The basic concept of diabatic CAES involves storage of off-peak energy for later use with only the fraction of energy (gas or oil) that would be used by a standard gas or oil turbine. It is the least efficient option in comparison with other systems. However, it is the most matured one and there are currently two plants in operation using this concept: the 321 MW Huntorf plant in northern-west Germany belonging to E.ON Kraftwerke and the 110 MW plant in McIntosh (Alabama, USA) belonging to PowerSouth company. Both facilities use underground salt caverns for air storage. Comparison of technical parameters of operating D-CAES plants is presented in Table 1. During peak-off hours, the electric driven compressor injects air into underground cavern. When the power is necessary for the grid, the process is reversed and the air is returned to the surface to be burned in a natural gas turbine. Combustion gas is expanded in the two-stage turbine to spin the generator and produce electricity. The cycle efficiency of D-CAES plants is 42% and 54% respectively for Huntorf and McIntosh. The difference between the efficiencies of the installations is due to the presence of a recuperator in the McIntosh plant, which additionally preheats the combustion air for the gas turbines.

Table 1. Comparison of technical parameters of operating D-CAES plants [8].

Plant	Huntorf	McIntosh
Cycle efficiency	42%	54%
Energy input for 1 kW h_{el} energy output	0.8 kW h_{el}/1.6 kW h_{gas}	0.69 h_{el}/ 1.6 kW h_{gas}
Energy content (in relation to power output)	642 MW h	2640 MW h
Planning-construction-commissioning	1969 - 1978	1988-1991
Compression		
Max. electricity input	60 MW	50 MW
Max. air mass flow rate	108 kg/s	Approx. 90 kg/s
Compressor units	2	4
Charging time	Approx. 8 h	Approx. 38 h
Storage		
Cavern pressure range	4.6 – 7.2 MPa	4.6 – 7.5 MPa
Cavern volume	310 000 m^3	538 000 m^3
Expansion		
Max. electricity output	321 MW	110 MW
Control range (output)	100-321 MW	10-110 MW
Discharging time (at full load)	Approx. 2 h	Approx. 24 h
Start-up time (normal/emergency)	14/8 min	12/7 min
Max. mass flow rate	455 kg/s	154 kg/s
HP turbine inlet	4.13 MPa/490°C	4.2 MPa/538°C
ND turbine inlet	1.28 MPa/945°C	1.5 MPa/871°C

The major problem in conventional diabatic CAES is the low efficiency as a result of heat loss during the compression stage. A number of concepts have been developed in order to preserve the heat and reuse it during the discharge process. Concepts which include heat loss in the process are called adiabatic CAES (A-CAES) and can be divided into A-CAES without and with separate Thermal Energy Storage. The first method is to store hot air inside the compressed air cavern (A-CAES without Thermal Energy Storage). The major drawback of this concept is the fact that adiabatically compressed air (up to a moderate 1 MPa) heats up to the temperature of 277°C and most of the CAS cannot withstand such high temperatures. Therefore, a development of this concept is to provide an additional, separate Thermal Energy Storage in order to store heat generated during the compression phase and add it when the air is decompressed. A more promising technology is the A-CAES with a TES (Thermal Storage System). In this case, the recovered process heat from the compression is stored in a separate TES and cooled air can be stored in a conventional Compressed Air Storage (CAS) [9].

The third concept, which is being under investigation, is the isothermal CAES where the increase of temperature is avoided when air is compressed, and decrease in temperature when the reservoir is discharged. For that purpose, a piston machinery is used which can perform a slow compression or expansion process when enough time is left for the heat exchange process inside the machinery. In this case, additional exchange contact surface area is needed or a liquid piston is used. Other concepts involve spraying liquids into the plug room of a common piston machine or the compression of a pre-mixed foam [8].

Among all the CAES methods, typical diabatic CAES is a proven and relatively effective method to store energy. The most important problem in this technology is the Compressed Air Storage which as shown in Table 1 has to be under high pressure (up to 7.5 MPa).

3. Underground Compressed Air Storage

The technology of storing air in caverns can be safe and reliable as proven for over 30 years by many natural gas storage facilities and two CAES plants in operation. The most common, proven and reliable location for CAS are salt caverns ie. solution mined salt domes. Salt caverns are widely used for natural gas storage and currently in Europe there are over 141 storage facilities accounting for over 98 168 Mm^3 of natural gas storage [10]. Unfortunately, the distribution of Permian salt domes which are favorable for natural gas storage or CAS is uneven. Majority of Permian deposits is located in the Northern part of Europe not necessarily close to energy sources, see figure 3 [11]. Other locations include Mesozoic (Spain, Portugal) and Tertiary (France, Romania) salt deposits.

As it can be seen in figure 3, the distribution of salt deposits limits the use of them as potential air storage. There are however other options where CAE can be stored and these are:

- Depleted gas reservoirs,
- Rock caverns,
- Small underground aquifers,
- Abandoned coal mines.

Depleted gas reservoirs could serve as CAS as the geology of such reservoirs is well known and the overburden is impermeable. However, the residual gas, remaining water and various oxide gases may cause significant problems to the surface installation (oxidation from the air, corrosion, etc.) in this case the deliverability of gas. Further, complex reservoir geology may disturb cyclic injection and production. So far, there is no installation for CAES in depleted gas reservoirs.

Another proposal is to store compressed air in aquifers. This concept is being investigated, but it suffers from many drawbacks, such as pressure drop by the loss of oxygen in the aquifer, microbial growth, and decrease in permeability of the reservoir [12].

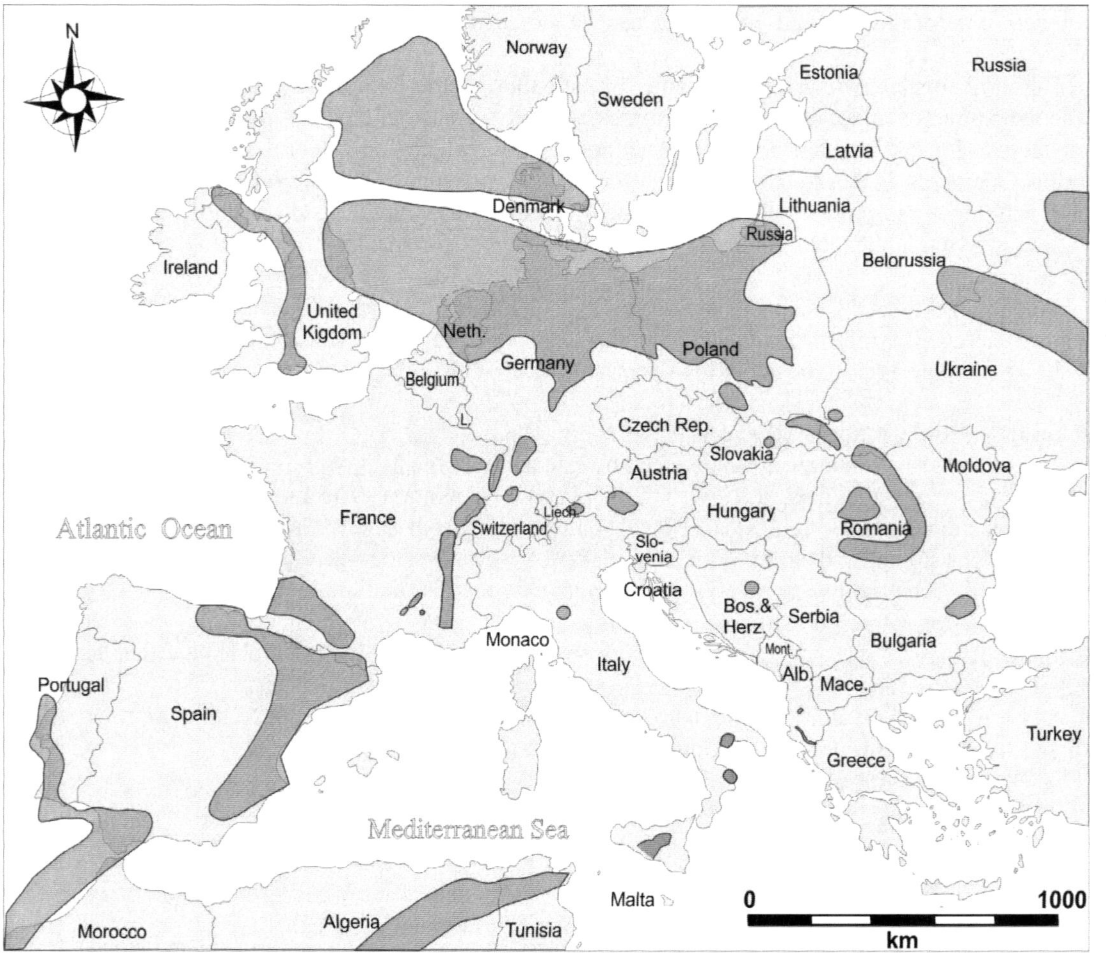

Figure 3. Salt deposits location in Europe [11].

Rock caverns excavated in hard rocks also have been used for liquid storage all over the world. This technique has two variants. Either the liquid is stored in unlined caverns using hydrodynamic sealing or the cavern is lined with metal and/or shotcrete/concrete [13, 14]. The first option is rather difficult to apply for CAS, however as an alternative, polymers are proposed by Terashita et al. [15] in a former Japanese zinc mine in Kamioka and a coal mine in Sunagawa.

Abandoned coal, rock and base metal mines could serve as potential candidates that can be converted into CAS. A huge advantage of coal mine location is the fact that in a majority of the cases they are located near power plants and reduces the energy transport lines. The concept of storing natural gas and CO_2 in abandoned coal mines is not new and for natural gas storage was proven in various locations [16, 17]. Three industrial scale installations are known and commercial operation started as early as in 1961. Two coal mines were used as temporary natural gas storage in Belgium: Peronnes and Anderlues, located between Mons and Charleroi and a third mine was in Leyden, USA. These installations prove that storing gas in coal mines is a feasible concept. It has to be highlighted that in case of abandoned coal mines natural gas storage can be enhanced by sorption on remaining coal seams. The Anderlues storage facility was operating between 1980 and 2000 at very low pressure (max 0.35 MPa). However, the reservoir volume below the thrust fault at 600 m depth was estimated by the operator between 6 and 10 Mm^3 (mainly based on regional assessments of accessible parts of old mine workings). This allowed for the storage of about 20 Mm^3 of CH_4 (14 000 tons) at working reservoir pressures. The amount of adsorbed CH_4 in coal was almost 8 times higher than the free gas capacity since the total storage capacity of the reservoir was 180 Mm^3 or 130 000 tons CH4. Sorption phenomena on remaining coal seams was also observed in gas reservoirs in Leyden and other Belgium mine [18]. This however cannot be the case in Compressed Air Storage since oxygen may cause self-combustion of coal and as a result an underground fire.

4. CAS in abandoned coal mines

As mentioned in the previous paragraph there is a possibility to store compressed air in abandoned coal mines if the drifts and shafts are properly sealed and separated from the remaining coal seams. The general concept of CAS in abandoned coal mine is presented in figure 4. This concept considers sealing of underground workings from the remaining coal seams and drifts which may be prone to collapse or deformation. It also considers applying liners on the surface of drifts and shaft in contact with air to prevent leakages.

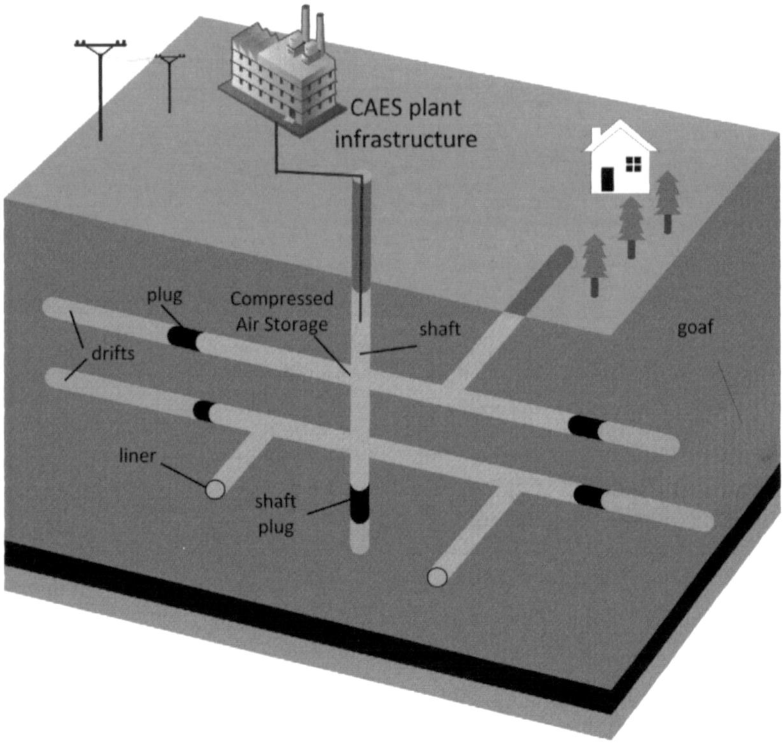

Figure 4. General concept of Compressed Air Energy Storage in abandoned coal mine.

Storage of air in closed/abandoned mines has some major advantages over surface storage in tanks and other locations, these are:

- Very high storage capacities of mines as CAS,
- Relatively low cost since the excavations already exist and the geology of the site is relatively well recognized,
- Many locations are directly connected to regional renewable energy sources or other energy sources,
- Safe operation and protection against external factors (weather, equipment failure, terrorist attacks) since the underground system is separated from the surface and connected only by valves and pipes,

- Large surface area available for the installation of the infrastructure (former mine surface facilities),
- Distribution of abandoned coal mines in urbanized areas in Europe where energy storage is a major problem.

A rough estimate of available volumes in an example coal mine located in the Upper Silesian Coal Basin in Poland is presented in table 2.

Table 2. Example calculation of coal mine volume

Type of workings	No. of workings	Volume (m^3)
Shafts	2	64 828.44
Level III (1050 m)	6	6 927.70
Winzes connecting levels III and II	1	26 805.60
Level II (850 m)	94	496 873.60
Winzes connecting levels I and III	20	153 210.00
Level I (650 m)	57	503 442.10
Total	180	1 252 087.44

Example calculation of volume in an typical coal mine located in the Upper Silesian Coal Basin in Poland shows that such volume is almost twice or three times larger than the volume of caverns used in existing CAES in Huntorf and McIntosh plants. Even when considering that large of parts or levels in a mine cannot be used for CAS, the remaining volumes would be still considerably high. This implies that the concept of CAES in closed or abandoned coal mines might be worth exploring. Obviously, there are still considerable problems to be overcome before this concept would be feasible.

The following, non-exhaustive, list of limitations related to this concept is shown below:

- Compressed air in contact with coal may cause spontaneous combustion of coal and as a result uncontrollable fire – all the remaining coal has to be separated from the storage site,
- Former mining operation at various depths causes cracks/fissures and fractures in the over- and interlaying strata or subsidence which may threaten overburden integrity and as a result will cause air leakage or infiltration of water into the compressed air storage,
- Coal mine flooding and uncontrollable inflow of water may reduce available volumes and increase moisture content of air.

Therefore, in order to start with any concept related to CAS in abandoned coal mines the following conditions have to be fulfilled:

- Storage facility has to be separated from any other coal mine to prevent air leakages or contacts with coal,
- Overburden integrity and drift stability is a major issue and should be carefully verified before any operation is started,
- Inflow of water has to be either controlled or location should be in dry mines,
- Workings accessible for air should be isolated with liners and plugs designed to withstand high pressures in storage facility (up to approximately 8 MPa).

Well recognized geological setting and information on the geomechanical properties of the rock mass surrounding the facility and choice of best available technologies of liners and plugs might ensure feasibility of this concept.

5. Conclusions

Increase in the use of renewables in the energy mix of European Union, apart from the obvious benefits, also creates problems for the grid operators. One of the solutions to deal with intermittency, unpredictability and local availability of renewables is the energy storage. So far, large scale energy storage sites are either pumped-storage hydroelectricity plants or Compressed Air Energy Storage facilities. In this paper the second concept is presented where particular focus is made on the underground storage part. Limited availability of salt caverns in Europe encourages researchers to seek for other options such as compressed air storage in closed or abandoned underground coal mines. This concept has distinct benefits such as close location to energy sources and relatively large volume of storage. Nevertheless, feasible studies are still immature and certain problems have to be overcome in order to develop a fully functional, large-scale operating concept. The major problems are related to the safety of operation i.e. isolation of workings designated for air storage from the remaining coal, provide leak tightness at high pressure, underground storage integrity and stability during injection/withdrawal cycles. Only after overcoming and solving these major issues this concept might be feasible.

6. References

[1] European Commission 2017 Eurostat. Europe 2020 indicators - climate change and energy
[2] Pérez-Arriaga I J 2011 Managing large scale penetration of intermittent renewables *2011 MITEI Symposium,* Framework Paper
[3] Brijs T, De Vos K, De Jonghe C and Belmans R 2015 *Renew. Energy* **80** pp 53–60
[4] Karki S U, Halbhavi S B and Kulkarni S G 2014 *Int. J. Adv. Res. Electr. Electron. Instrum. Energy* **2014**
[5] The Fuel Cells and Hydrogen Joint Undertaking 2015 Study on commercialisation of energy storage in European market to 2030 and beyond *Fuel Cells Bull.* **4** pp 14–6
[6] Gay F W 1948 Means for storing fluids for power generation
[7] Deutsche Energie-Agentur GmbH 2012 *Druckluftsysteme in Industrie und Gewerbe: Ein Ratgeber zur systematischen energetischen Modernisierung*, Berlin 32
[8] Budt M, Wolf D, Span R and Yan J 2016 *Appl. Energy* **170** pp 250–68
[9] Barbour E, Mignard D, Ding Y and Li Y 2015 *Appl. Energy* **155** pp 804–15
[10] European Commission. Directorate-General for Energy Internal Energy Market 2015 *The role of gas storage in internal market and in ensuring security of supply*
[11] Gillhaus A and Horvath P-L 2007 *Compilation of geological and geotechnical data of worldwide domal salt deposits and domal salt cavern fields* (Clarks Summit (PA): Solution Mining Research Institute)
[12] Letcher T M 2016 *Storing Energy: with Special Reference to Renewable Energy Sources* (Elsevier)
[13] Kim H-M, Lettry Y, Ryu D-W and Song W-K 2014 *Mater. Struct. Constr.* **47** pp 127–40
[14] Kim H-M, Rutqvist J, Kim H, Park D, Ryu D-W and Park E-S 2015 *Rock Mech. Rock Eng.* **49** pp 573–84
[15] Terashita F, Takagi S, Kohjiya S and Naito Y 2005 *J. Appl. Polym. Sci.* **95** pp 173–177
[16] Piessens K and Dusar M 2004 *Geol. Belg.* **7-3/4** pp 165-180
[17] Lutynski M 2010 *Gospod. Surowcami Miner.-Miner. Resour. Manag.* **26** pp 93–104
[18] Jalili P, Saydam S and Cinar Y 2011 *Coal Oper. Conf.*

IOP Conf. Series: Materials Science and Engineering **268** (2017) 012007 doi:10.1088/1757-899X/268/1/012007

The determination of methane resources from liquidated coal mines

Stanisław Trenczek

Institute of Innovative Technologies EMAG, Leopolda 31, 40-189 Katowice, Poland

Stanislaw.Trenczek@ibemag.pl

Abstract. The article refers to methane presented in hard coal seams, which may pose a serious risk to workers, as evidenced by examples of incidents, and may also be a high energy source. That second issue concerns the possibility of obtaining methane from liquidated coal mines. There is discussed the current methodology for determination of methane resources from hard coal deposits. Methods of assessing methane emissions from hard coal deposits are given, including the degree of rock mass fracture, which is affected and not affected by mining. Additional criteria for methane recovery from the methane deposit are discussed by one example (of many types) of methane power generation equipment in the context of the estimation of potential viable resources. Finally, the concept of "methane resource exploitation from coal mine" refers to the potential for exploitation of the resource and the acquisition of methane for business purposes.

1. Introduction

Methane in the nature occurs mainly in natural gas deposits – as its primary ingredient – and in hard coal deposits – as adsorbed gas. In the first case, it is the sought after and desirable high-energy fuel, in the second, after release, poses a serious threat to people and the environment in which methane-air mixture may form.

Coal deposits in the Upper Silesian Coal Basin are, apart from its northern part, strongly methane-bearing. By the mid-1950s, methane found in underground mines was not a significant problem, although it was noted in the history of the Polish mining industry by such catastrophic incidents as those in the mines: Brzeszcze-Silesia (10.06.1945 – 9 victims, Marcel (06.11.1947 – 5 victims), Mysłowice (4.12.1951 – 7 victims), methane explosion and coal dust in Jankowice mine (04.05.1950 – 29 victims).

Beginning in the 1950s the mining works, which prepared the exploitation of the southern part of the GZW in the Rybnik Coal Region, as well as the exploitation of ever deeper coal seams in the remaining part of the GZW, caused that the methane threat, or more precisely, the methane explosion hazard became a very common occurrence (outside the northern part of the GZW). This is evidenced by such mine disasters as: [1]: methane explosions in mines Moszczenica (08.04.1959 – 14 fatalities), Silesia (28.06.1974 – 34), Halemba (10.01.1990 – 9 victims), Śląsk (18th of December 1990 – 4), Sośnica (07.11.2003 – 3), Borynia (04.06.2008 – 6), Wujek Ruch (06.10.2014 – 5), or methane and coal dust explosions in mines: Halemba (21.11.2006 – 23) and Wesoła (13.01.2008 – 2). There were also methane and rocks' outbursts in Zofiówka (22.11.2005 – 3).

Combating methane hazard until the seventies of the last century practically consisted in ventilations works. In the following years, after the previous tests, the use of methane drainage has become more and more often used to improve safety, which over time has become mandatory for use under certain conditions [2]. This was treated as a significant step in the fight against the explosion, which indeed was, and still is, a great success despite the disasters associated with methane. It should be emphasized that less methane flows in the mine air, posing a risk of explosion (other than places designated for this purpose), as methane can be captured. However, methane was not treated for a long time as a fuel – it was either discharged into the ventilation air – even underground, or directly into the atmosphere – from the surface methane pump station.

It was only in the 21st century that methane derived from hard coal was started as a valuable fuel and a factor which improves the economics of coal production [3]. Methane removed by methane drainage method is used as a low-methane fuel in various types of heating and power plants, eg in boiler houses with gas or coal-fired boilers and with gas turbine engines and turbines, or in gas-fired boilers. Following this, in the context of an ever-increasing number of liquidated mines, including methane mines, it was thought to take methane of such former mines and also about its economic use in heating and power plants. It just started to treat it as a business idea, which forced the alignment of legislation to document methane deposits from hard coal deposits as a primary mineral, as well as the granting and succession of concessions to entrepreneurs/mines, which are transferred to the structures that are involved in the restructuring of mines.

It is worth to note the data from the Ministry of the Environment (for example: [4]), which indicates that the size of the documented methane resource reserves in the Upper Silesian Basin is estimated at over 90 billion cubic meters. in 60 deposits. These resources are increasing, as more and more reconnaissance work is being done, resulting in new documentation of methane deposits from hard coal deposits. Only in 2015, the stock of methane from hard coal deposits increased by nearly 4 billion cubic metres, in the connection with documenting new three deposits. There is also an increase in the size of industrial reserves, which in the case of 27 deposits fluctuated within 6 billion cubic meters. At the end of May 2017, the searching for and the exploration of methane deposits from coal seams was conducted in Poland on the basis of eight concessions granted by the Ministry of Natural Resources and the extraction of this gas – on the basis of three concessions.

2. Methane in hard coal deposits

Methane in the coal deposits is divided into the accompanying mineral – the methane exploitation is possible during coal mining - and the main mineral that can be extracted either independently of the coal or in the post-mining period, ie exploiting the methane bed from the hard coal mines. According to the subject of this article, the next part of this article will cover the last case.

The occuring of methane in coal, as it is commonly accepted, is associated with the formation of coal. In low carbon coals methane was formed as a result of microbial activity, and in coals with a higher degree of carbonation – as a result of the organic matter processes [5] accumulated in the marshes as a residue of various vegetation growing on Earth in hot and compressed climates. The chemical and physical reactions that have caused them have led to the formation of coal, as well as methane, carbon dioxide, nitrogen and water. As the deposits of organic matter exploded, the pressure and temperature increased, resulting in an increased degree of carbonation of the material and an increase in the methane content.

It is characteristic for methane, found in coal seams, that it contains small amounts of carbon dioxide and nitrogen, and does not contain hydrogen sulphide (so methane is so-called "sweet" gas). This means that methane does not require special preparation before use. In addition, it is extracted at a much lower pressure than methane from conventional deposits.

Methane contained in coal [6, 7] is adsorbed (in micropores less than 2 mm in diameter), in form of free gas trapped in gaps and cracks and dissolved in water in coal. In the carbon matrix it is trapped by weak van der Waals bonds, characterized by the attraction between noble gases and between non-polar molecules.

As practice shows, methane in coal occurs in varying amounts, and the amount of sorbed methane depends on many factors. The main ones are the sorption properties of coal, the temperature and the

IOP Conf. Series: Materials Science and Engineering **268** (2017) 012007 doi:10.1088/1757-899X/268/1/012007

deposit pressure. Accumulation of methane in the seam promotes an increase in carbonation and a rise in pressure, with a decrease in humidity and a temperature of coal.

Coal has a specific sorbent capacity for certain values: pressure, temperature and humidity. The amount of hydrocarbons that can be absorbed results inter alia from petrographic composition and degree of coagulation [8, 9]. However, due to the complexity of the problem, there is no clear correlation between methane content and the factors that form it. The reason for this is that some of the organic matter has evaporated in the carbonizing process, and some have evaporated in the areas of tectonic disturbances. Disturbances can, in certain circumstances, be a convenient way of migration, and in others create sealed screens that prevent degasification.

The phenomenon of adsorption occurring on the solid-gas phase boundary is a big issue, so the theory on the subject is a lot. One of the most commonly used models to describe sorption processes is the Langmuir model, where the amount of adsorbed methane in coal is characterized by the equation [5]:

$$\frac{p}{a} = \frac{1}{K \cdot a_m} + \frac{p}{a_m} \tag{1}$$

where:
p – equilibrium pressure [MPa],
a – amount of adsorbed methane per unit mass of coal [m^3/Mg],
a_m – the amount of adsorbed methane assuming the one-molecule layer [m^3/Mg],
K – adsorption equilibrium constant, determined experimentally.

An example of the course of changes in the amount of adsorbed methane in coal, as a function of temperature change – with constant moisture content, light fractions, pure carbon and ash – is shown in figure 1.

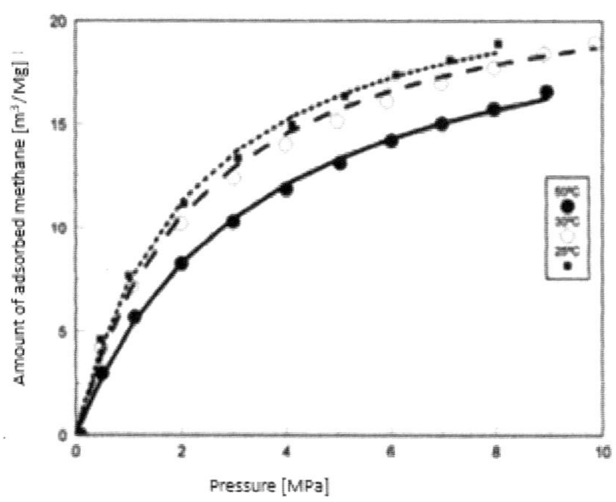

Figure 1. The examples of methane adsorption isotherms in coal for different temperatures, with a constant content of: moisture – 1.4%, light fractions – 24%, pure carbon – 67%, ash – 7.5% [5].

In practice, the amount of gas (including methane) G [m^3] contained in the coal seam is calculated from the dependence:

$$G = A \cdot h \cdot r_c \cdot G_s, \tag{2}$$

where:
A – surface of coal seam [m^2],
h – average thickness of the coal seam [m],
r_c – average coal density [Mg/m^3],
G_s – amount of adsorbed methane [m^3/Mg].

In turn, the amount of adsorbed methane in the coal mass unit G_s [m^3] is calculated as [5]:

$$G_s = (1 - f_a - f_m)\frac{V_L \cdot P}{P_L + P}, \tag{3}$$

where:
f_a – percentage of ash [%],
f_m – percentage of moisture [%],
V_L – volume of Langmuir [m^3],
P_L – Langmuir pressure [MPa],
P – pressure [MPa].

As it turns out, the determination of the amount of methane that can be adsorbed from the coal seam is the more accurate when the well-known values of the parameters are taken into account.

3. Factors characterizing the methane deposit from coal deposits of liquidated mines

As it is already mentioned, the coals show a certain sorption capacity. Under given conditions of pressure, temperature and humidity, they can absorb the right amount of hydrocarbons, resulting from, among other things, the petrographic composition and degree of carbonation. As these processes are very complex, one can not expect a clear correlation between methane content and the factors that form it. Significant amount of gas produced by the process of carbonation of the organic matter has evaporated, and therefore the conditions for re-generation of the gas have to occur. Probably methane migrating from a deeper subsoil was absorbed by the coals to which it had come.

The methane bed found in the coal deposits of the liquidated mine is:

- free methane, which filled post-mining goaf, non-caved areas and/or not in bad repair excavations, as well as pores and gaps in coal seams and porous barren rock;
- methane sorbed, that is physico-chemically bound to coal mineral matter of coal seams and dispersed in barren rocks.

The methane drainage from such a deposit usually occurs when the developed business plan confirms the profitability of the project. The business plan is usually based on the geological and investment documentation of the methane deposit in the hard coal deposit, which must be prepared on the basis of the geological and investment documentation of the hydrocarbon reservoir [10]. For its preparation, quite extensive knowledge is needed – mainly in the field of geology, including:

- hard coal deposits in the given area, their range of methane content, porosity, ash content, humidity or density,
- barren rocks deposits in strata, mainly relating to their porosity and methane content,
- the structure of mining excavations,
- the method of development of coal deposits in individual seams,
- the coal mining made in individual seams,
- the rock tectonics, including downthrows, upthrows, faults,
- the vicinity of the deposit, which may affect the inflow or outflow of free methane, water, other gases,
- methane depletion in the period from the preparation of the coal deposit to exploitation, through the period of mining plant operation until its physical liquidation – including depletion by ventilation or methane drainage,
- the parameters of free gas in the free space of the deposit.

Practice shows that the knowledge of hard coal deposits in a given area is relatively accurate, as the documentation of hard coal deposits has been based on the results of research on coal and rock samples, in terms of methane content, porosity, moisture, and coal samples also in terms of ash content and density. In turn, the development of the deposit and the exploitation of the given coal seam based on the specific structure of the excavation knowledge has deepened and expanded due to ie. rock tectonics.

The formation of soil or the neighborhood of a given deposit may not be accurately identified – an accurate and extensive knowledge of the neighboring mining area is not always known. It is also difficult to accurately determine the amount of methane loss in the entire earlier desorption period and therefore it is estimated. The knowledge of the parameters of gases creating a free gas is the most accurate, as it results from the in situ research needed to carry out a concession for the exploration and discovery of methane deposits from coal deposits.

In the abandoned mines, the original methane content system of the coal seams has changed many times. The sorption and desorption processes were repeated with the coal mining, which was usually accompanied by the methane drainage, which at the end of mining formed a new spatial distribution of methane content. Its role in this is the mutual arrangement of a *sorbed gas - a free gas* that determines many of the physico-chemical factors arising, inter alia, from geological conditions, as well as from the migration of gases outside the given area (eg to the surface, adjacent areas with active mining, etc.). Therefore the fall of the deposit pressure causes the gas desorption from the seams to cracked and porous rocks, goaf, and especially to the post-excavation workings.

Due to the fact that in Poland – in a decisive part – it is made exploitation of coal seams by longwall working system, in the rock there are formed large cavities that are the residues after the extracted coal - the goaf. The caving zone (figure 2) includes approximately 2.5 to 3.5 times the extracted coal seam [11] (h_b – in figure 2), and the high caving zone includes to 5 times [12] ($h_b + h_z$ – in figure 2) and even up to 6-7 times [13, 14, 15]. The area around the longwall face is characterized by very high values of filtration and permeability coefficient.

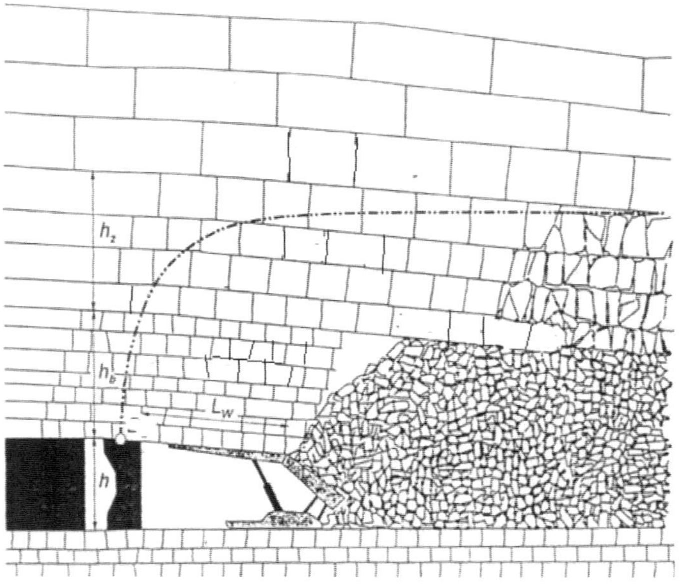

Figure 2. A diagram of the relaxed rock mass around the caving longwall face [12].

Another important factor for the methane deposit is the convergence of both goaf and mining excavations: close to longwall, air workings and rock drifts [12].

4. Identification of methane resources from hard coal deposits

The methane bed of the hard coal mine requires, as already mentioned, documentation of what is being done during exploration-identification work. The scope of such work is usually specified in the concession for the exploration and identification of methane from hard coal. Once completed, appropriate documentation shall be prepared in accordance with the Geological and Mining Law [16] and the Implementing Regulations [17, 18, 19, 20, 10, 21, 22].

On the other hand, the classification categories for the deposit are based on the qualification rules for the documentation of hard coal methane, as set out in the Balancing Criteria and the Principles for Calculation of Methane Resources in Coal Seams [23] and the Principles for Documenting Oil, Natural Gas and Methane in the seams of coal [22]. Accordingly, the methane deposit, as primary mineral, is eligible depending on the surface area for which one drainage hole is made, in which methane content tests were performed, which are required to determine the industrial resources, ie:

- in category C if at least 1 hole/8 km^2,
- in category B, at least 1 hole/2 km^2,
- to category A, includes the methane resources documented by exploited exploitation holes using the mass balance method or statistical methods.

It is also important to determine the methane boundaries of the given area, which, in addition to, as already mentioned, account for the presence of methane sorbed in the bed and free methane in the free space must take into account the methane in the barren rocks. As the lower limit is usually the level (depth) of rock deposits, and the upper limit for the occurrence of methane sorbed on carbon deposits is usually the Carboniferous roof, located below the impermeable forms of Miocene clay.

In the case of liquidated coal mines where coal remains unextracted, it is important to identify the methane deposit from this part of the coal bed. This is done in terms of balance resources and off-balance resources. The balance resources consist of sorbable methane, which is physicochemical-bound to carbonaceous material and dispersed in barren rocks, and methane free that fills post-mining goaf, free spaces in narrow workings, as well as pores and cracks in coal seams and porous rocks. Off-balance resources consist of non-sorbable methane.

Significant, from the point of view of the future methane drainage from the liquidated mine, is to determine the size of the dynamic resources of methane sorbed on the coal seams, which are partly a potential source of methane to the open space - eg non-liquidated excavations, goaf. It is a sorbed methane, which is released from a part of the rock covered by previous mining works in the area and, if so, carried out in the area of the adjacent coal bed. Also important is the recognition of the tectonics of the rock, because in the tectonic region strongly disturbed is the diversity of methane content, even within one seam. There are also - in certain conditions - the possibility of methane inflow even from more distant areas and deposits through these disturbances, and in others - they can create sealed screens that prevent degassing.

Depending on the specific methane conditions prevailing in the coal bed methane and in the goaf, an appropriate method of calculating the methane content of coal in the reservoir is chosen. Typically, aggregate resources of methane from coal in the deposit (*Qm*) are composed of static methane free (*Qg*) resources filling the anthropogenic reservoir and the dynamic resources of the sorbed methane (*Qd*) in the hard coal deposits formed to the depth assumed as the bottom boundary:

$$Qm = Qg + Qd \ [\mathrm{Nm^3}] \tag{4}$$

4.1. Static resources of methane deposits

Assessment of the amount of static resources of free methane (Qg) accumulated in the anthropogenic reservoir – collector created in the rock cavities – what are post-mining gobs and dammed workings, including main workings, non-liquidated workings, made in the barren rock (left as full-scale but converged) and workings made in coal deposit (left in reduced dimensions and converged), workings partially liquidated (eg by stowing) by volume method, using the formula:

$$Qg = Vk \cdot p \cdot C \cdot wT \ [\mathrm{Nm^3}] \tag{5}$$

where:
Vk – volume of collector within free gas accumulation [m^3],

p – the gas pressure coefficient determining the deposit pressure relative to the normal pressure,
C – content - average – CH_4 in gas [%],
wT – conversion factor from m^3 to Nm^3 (Nm^3 – normal m^3, for normal conditions, ie pressure 101,325 kPa, temperature 273,15 K, coefficient of compressibility 1).

The volume of the collector is calculated as the sum of the individual available gas volumes for the free space - ie non-liquidated excavations (Vw_x) and goaf:

$$Vk = Vw_{sk} + Vw_w + Vw_p + Vzr \ [\mathrm{m}^3] \qquad (6)$$

where:
Vw_{sk} – volume of non-liquidated excavations made in the barren rock [m^3],
Vw_w – volume of non-liquidated excavations made in the coal seam [m^3],
Vw_p – volume of non-liquidated workings close to longwall made in the coal seam [m^3],
Vw_{zr} – volume of post-mining goaf [m^3],

The volume of non-liquidated workings Vw_x is calculated as:

$$Vw_x = L \cdot A \cdot C_x \ [\mathrm{m}^3] \qquad (7)$$

where:
L – the length of non-liquidated excavations of a given type - separately: excavations made in the L_{sk} barren rock, excavations made in the L_w coal seam, excavations made in the coal seam close to longwall L_p) [m],
A – cross section of the narrow working corridor of a given type - separately: excavations A_{sk} made in the barren rock, excavations made in the coal seam A_w, excavation made in the coal seam close to longwall A_p) [m^2],
c_x – capacity coefficient of free space for a given type of excavation, assuming that mine work has been retreated for a minimum of 20 years back – separately: for excavations made in barren rock c_{sk} = 0.8; for excavations made in the coal seam c_w = 0.5; excavation made in the coal seam close to longwall c_p = 0.3).

On the other hand, the volume of goaf is calculated according to the formula:

$$Vzr = \sum \left(P_x \cdot g_{px} \cdot c_{zr} \right) \ [\mathrm{m}^3] \qquad (8)$$

where:
P_x – the surface of each extracted seam [m^2],
g_{px} – average thickness of each exploited seam [m],
c_{zr} = 0.15 – the capacity factor of the goaf, assuming that mine work has been retired a minimum of 20 years back.

4.2 Dynamic resources of methane deposits

The amount of sorbed methane (*Qd*) is taken for the coal deposits from the Carboniferous roof to the depth assumed as the limit. The methodology most commonly used is:
- estimation of pure coal mineral matter resources (csw) at individual levels in the separated parts of coal deposit,
- calculation of the mean methane content of the seams in those parts, which is calculated from the results of the tests at the different levels of the separated part of the deposit,
- multiplying the amount of pure coal mineral matter resources (csw) by the average methane content minus the residual methane content value, which is calculated by indirect analytical method, due to lack of direct measurements. In csw estimation, documented coal resources are corrected by

multiplying it by the increase factor for undocumented, thin seams and coal inserts with a thickness of 0.1 m and a reduction factor for moisture and ash content in coal.

Generally, the assessment of the dynamic resources of methane in the bed (*Qd*) is made using the formula:

$$Qd = Qw \cdot K1 \cdot K2 \cdot (G - Gr)\ [\mathrm{m}^3] \quad (9)$$

where:
Qw – documented coal resources [Mg],
G – average methane content [m^3 CH_4/Mgcsw],
K1 – increase factor for undocumented coal inserts (depending on their total thickness - usually within the range of 3.0-6.0),
K2 – reduction coefficient due to moisture and coal ash (usually within the range of 0.75-0.85),
Gr – residual methane content [m^3 CH_4/Mgcsw], which is calculated for coals using an analytical method for the determination of residual (non-desorbable) regression equation:

$$Gr = a_0 + a_1\,\Delta P_2 \quad (10)$$

where: a_0, a_1 – coefficients of regression equation.

5. Limitations of methane drainage potential

As already mentioned, in the second decade of the 21st century, the exploitation of methane from the hard coal deposits from liquidated mines was increased. Most methane is used to produce electricity in cogeneration engines, to which methane is supplied from pumping stations. The pumping station sucks gases, including methane, from the methane bed through a borehole made from the surface to an anthropogenic reservoir (usually to the goaf of the just extracted top seam; below the top seam there are situated the gobs of bottom seams) and transfers them to the cogeneration engine.

The methane mining from hard coal to dry up completely is not possible. The first reason is the limitations that result from the parameters of pumping stations. This concerns the vacuum that could be generated. The amount of vacuum created (P_{Pgr}), depending on the operating parameters of the device used to drain the methane from the coal bed is usually the pressure level of approx. P_{Pgr} = -35 kPa. This is the limit pressure for the emission of desorbable gas. At such pressure, the equilibrium between the pressure of the gas in the coal deposit and the vacuum produced and the possibility of further methane operation are terminated. These quantities of methane remaining in the deposit are non-industrial resources (Q_{np}), which together with industrial resources (Q_p) are the total methane resources of hard coal methane (Q_m).

The second reason for limiting the potential for full exploitation of methane resources is the need to ensure the safe operation of the cogeneration engine.

For example, the Caterpillar 3516 1.136 MW engine for optimum operation and production of electric current requires 4.7 m^3 of CH_4 at a concentration of 100%. Engine operation is possible with a minimum methane concentration in the engine gas of Cs_{min} = 33% CH_4. When the concentration drops below this value, the motor is automatically immobilized. This means that at this limit concentration the gas should be delivered at 15 m^3/min.

These acts make it necessary, from a business point of view, to introduce the concept of "methane exploitation resources" from the hard coal mine – Q_E, which can be completely used to produce electricity. These resources are defined as

$$Q_E = Cs_{min} \cdot Qm \cdot 0.01\ [\mathrm{m}^3] \quad (11)$$

As a consequence, this reason is more important because it determines the amount of resources that can be exploited.

6. Summary

Methane from hard coal deposits can be of two kinds: typical, that is, creating a methane explosion hazard in active mines and less typical – using it for electricity production – where as an energy fuel increases its share in the energy balance.

Estimating methane resources in the hard coal deposit requires detailed analysis of the geological and operation documentation both from the period of coal exploration and mine exploration, particularly methane content research.

The use of cogeneration engines for methane combustion and electricity generation creates concrete limitations in the potential for complete exhaustion of industrial methane from hard coal deposits – the operation below the specified methane concentration in the combusted gas engine is not possible.

From the business point of view, it was necessary to take into account the limitations resulting from the performance of the cogeneration engine in determining the amount of resources to be used, which led to the introduction of the term "methane exploitation resources" from the hard coal mine deposits.

References

[1] Katastrofy górnicze w Polsce. https://pl.wikipedia.org/wiki/Katastrofy_g%C3%B3rnicze_w_Polsce

[2] Rozporządzenie Ministra Gospodarki z dnia 28 czerwca 2002 r. w sprawie bezpieczeństwa i higieny pracy, prowadzenia ruchu oraz specjalistycznego zabezpieczenia przeciwpożarowego w podziemnych zakładach górniczych (Dz. U. z 2002 r. Nr 139 poz. 1169)

[3] Gatnar K 2010 Energetyczne wykorzystanie metanu z pokładów węgla – doświadczenia JSW S.A. i perspektywy w aspekcie zmian w Prawie Energetycznym, *Zeszyty Naukowe Instytutu Gospodarki Surowcami Mineralnymi i Energią Polskiej Akademii Nauk* **78**

[4] Górnictwo: rośnie rola metanu jako źródła energii. W: http://nettg.pl/news/143100/ gornictwo-rosnie-rola-metanu-jako-zrodla-energii

[5] Aminian K 2009 Evaluation of Coalbed Methane Reservoirs. Petroleum & Natural Gas Engineering Department, West Virginia University

[6] Bumb A C and McKee C R 1988 Gas-Well Testing in the Presence of Desorption for Coalbed Methane and Devonian Shale 'SPE Formation Evaluation'

[7] Hagoort J 1988 *Fundamentals of Gas Reservoir Engineering*. Elsevier

[8] Chand Bansal R and Goyal M 2009 *Adsorpcja na węglu aktywnym* (Warszawa: WNT)

[9] Ościk J 1979 *Adsorpcja* (Warszawa: PWN)

[10] Rozporządzenie Ministra Środowiska z dnia 1 lipca 2015 r. w sprawie dokumentacji geologiczno-inwestycyjnej złoża węglowodorów (Dz. U. z 2015 r., poz. 968)

[11] Ropski S and Znański J 1965 Zachowanie się stropu nad pokładem wybieranym z zawałem stropu. *Przegląd Górniczy* **10** pp 3–12

[12] Biliński A and Kostyk T 1993 Wpływ czynników naturalnych i technicznych na stan utrzymania stropu wyrobiska ścianowego. *Prace Naukowe GIG* **779**

[13] Singh M M and Kendorski F S 1981 Strata disturbance prediction for mining beneath surface water and waste impoundments *1st Conference on Ground Control Mining* pp 76–89

[14] Palchic V 2003 Formation of fractured zones in overburden due to longwall mining *Environmental Geology* **44** pp 28–38

[15] Karacan C Ö 2009 Reconciling longwall gob gas reservoirs and venthole production performances using multiple rate drawdown well test analysis *International Journal of Coal Geology* **80** pp 181–95

[16] Ustawa z dnia 9 czerwca 2011 r. Prawo geologiczne i górnicze (Dz. U. z 2015, Poz. 196)

[17] Rozporządzenie Ministra Środowiska z dnia 22 grudnia 2011 r. w sprawie dokumentacji geologicznej złoża kopaliny (Dz. U. Nr 291, Poz. 1712)

[18] Rozporządzenie Ministra Środowiska z dnia 20 grudnia 2011 r. w sprawie korzystania z informacji geologicznej za wynagrodzeniem (Dz. U. Nr 292, poz. 1724 z późn. zm.)
[19] Rozporządzenie Ministra Środowiska z dnia 9 czerwca 2015 r. w sprawie przekazywania informacji z bieżącego dokumentowania przebiegu prac geologicznych (Dz. U. z 2015 r., poz. 903)
[20] Rozporządzenie Ministra Środowiska z dnia 1 lipca 2015 r. w sprawie dokumentacji geologicznej złoża kopaliny, z wyłączeniem złoża węglowodorów (Dz. U. z 2015 r., poz. 987)
[21] Rozporządzenie Ministra Środowiska z dnia 1 lipca 2015 r. zmieniające rozporządzenie w sprawie szczegółowych wymagań dotyczących projektów robót geologicznych, w tym robót, których wykonywanie wymaga uzyskania koncesji (Dz. U. z 2015 r. Poz. 964)
[22] Zasady dokumentowania złóż ropy naftowej, gazu ziemnego i metanu w pokładach węgla". Opracowanie zrealizowane pod kierunkiem Marka Niecia. Ministerstwo Środowiska. Departament Geologii i Koncesji Geologicznych. Komisja Zasobów Kopalin. *Agencja Wydawniczo-Konsultingowa „GEO"*, Warszawa 2002
[23] Kryteria bilansowości i zasady obliczania zasobów metanu w pokładach węgla. Ministerstwo Środowiska 1994

IOP Conf. Series: Materials Science and Engineering **268** (2017) 012008 doi:10.1088/1757-899X/268/1/012008

Costs assesment of aerological hazards prevention exemplified by a coal mine company

Dariusz Musioł

Faculty of Mining and Geology, Silesian University of Technology, Akademicka 2, 44-100 Gliwice, Poland

dariusz.musiol@polsl.pl

Abstract. A range of preventive works has been presented based on the example of one of the exploited longwalls and the costs borne by the coal mine have been exhibited. Further, the paper presents the costs of preventive aerological practices conducted for the improvement of work safety, borne by one of the coal mining companies. The paper also includes the calculation of cost indices. Based on these cost indices, an analysis of costs of the applied aerological preventive practices was conducted for the longwalls driven in years 2014-2015 by one of the coal mining companies.

1. Introduction

The exploitation of coal mine deposits in underground mines is becoming increasingly difficult and costly. The fixed and variable costs borne by the coal mines are mostly dependent on the amortization costs of machines and devices, the costs of materials and energy, remunerations, external services, taxes and fees as well as other prime costs [1, 2].

Conducting underground exploitation is inseparably linked to the occurrence of natural hazards. The group of aerological hazards includes methane hazard, spontaneous fire hazard, temperature hazard and coal dust explosion hazard. These hazards have a direct impact on the health and the life of the employees working in the mine. Due to the above, coal mines are obliged to conduct continuous preventive practices to limit the hazards occurring in the locations where the miners are working.

Preventive practices conducted against aerological hazards aimed at their limitation generate high costs, which are considered the variable costs of the conducted exploitation [3].

These hazards have a considerable impact on the total costs of a tonne of hard coal. The mean percentage of aerological preventive practices costs is within 4÷16.5% of the price of 1 Mg of coal [3]. As it may be noted, the necessity to conduct continuous preventive practices against aerological hazards to ensure functional safety [4] during exploitation is a cost that has a considerable impact on the price of coal.

The article includes the appraisal of costs of aerological prevention practices conducted in the coal mines of one of the coal mining companies [5]. The presented costs consider the aerological preventive practices applied for longwalls exploited in the years 2014-2015. The appraisal of the costs of the applied aerological preventive practices was based on the cost indices. The total costs borne within the applied practices were prepared based on adequately drawn up universal tables summarizing the individual costs occurring during the performance of applicable preventive works.

2. An example of a longwall – preventive works and their costs

The longwall [5, 6] was driven using a longitudinal caving system. The length of the wall was 230 m and the face height was 2.0 m. The longwall panel length was 1280 m. The ventilation was conducted using the 'Y' method from the boundaries (figure 1). All aerological hazards occurred in the longwall: methane hazard – category 4, fire hazard – 2nd group of spontaneous ignition susceptibility of coal, 3rd critical level of climatic hazard and coal dust explosion hazard class B.

The preventive practices against the methane hazard encompassed mainly the removal of methane from the ventilating road as well as the additional sealing of the side wall in the heading 1a following the advance of the longwall using foams, filling of the heading 1 using foams and the use of jetflow air movers for providing air flow. The prevention against the fire hazard was mostly aimed at the additional sealing of the pillar between the heading 1a and the ventilating road. The preventive works pertaining to the climatic hazard consisted in the cooling of air using group system cooling units in the area of the heading 1. The preventive practices pertaining to the coal dust explosion encompassed standard measures, that is, dust barriers and dusting zones.

The longwall was characterized by a relatively high mean output of 1 715 Mg/day. The exploitation period of the longwall was 563 days and the total output reached 965 532 Mg.

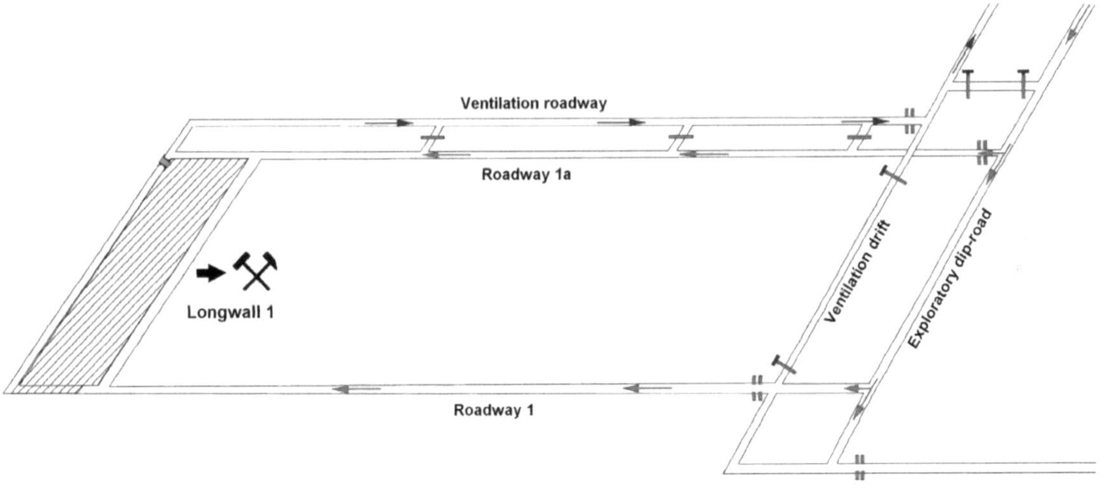

Figure 1. Layout of the longwall 1 [8].

The scope of works applied within the aerological preventive works in the longwall 1 including the unit costs has been presented in tables 1-4 [5, 7].

Table 1. List of costs of methane prevention practices for longwall 1 [5, 7].

No.	Name of the cost	Number/amount	Unit cost (zł)	Total cost (zł)
1.	Preparing drainage boreholes. Total No. of running meters of the drainage boreholes - 34 400 No. of boreholes - 316 No. of sets - 53	34 400	250 zł/1 running meter	8 600 000.00 zł
2.	Methane removal	19 months	94 210.00 zł/month	1 789 990.00 zł
3.	The performance of auxiliary ventilation devices (a – ventilating brattice lining material b – ventilating barriers, c – jetflow air movers, d – auxiliary fans, e – additional sealing of the caving with foams, f – insulating plugs etc.)	a- 0 b- 0 c- 2 pcs. d- 2 pcs. e- 359 895kg f- 14 pcs.	0 0 6 000 zł/piece 70 000 zł/piece 2.63 zł/kg 23 337 zł/piece	0 0 12 000.00 zł 140 000.00 zł 946 523.85 zł 326 718.00 zł
4.	Performance of a drainage roadway		Not applicable	
5.	Maintenance of the ventilation roadway in case of the Y-ventilation system	1 250 m	567.00 zł/1 running meter	708 750.00 zł
6.	Man-days related to preventing the methane hazard	11 403 man-days	547.19 zł/man-day	6 239 607.57 zł
7.	Energy costs WLE 1013B x 2 Huragan 120 jetflow air movers 2x – 19440 m^3/d x2	520 days x2 550 days x2	114.24 zł 625.97 zł	118 809.60 zł 688 567.00 zł
8.		**Total cost of the preventive practices**		**19 570 966.02 zł**

Table 2. List of costs of fire prevention practices for longwall 1 [5, 7].

No.	Name of the cost	No./Amount	Unit cost (zł)	Total cost (zł)
1.	Use of chemical materials: - light foams - heavy foams, - anti-pyrogenic agents - polyurethane cements	 0 dm^3 9 000 dm^3 0 dm3 1 300 dm^3	 0 zł/dm^3 5.40 zł/dm^3 0 zł/dm3 7.48 zł/dm^3	 - 48 600.00 zł - 9 724.00 zł
2.	Additional works for the application of chemical materials (boreholes, injection)	-	-	-
3.	Use of mineral materials: - mineral-cement binding materials, - cement, - sand	 0 Mg 15 Mg 50 Mg	 0 zł/kg 385.00 zł/Mg 22.96 zł/Mg	 - 5 775.00 zł 1 148.00 zł
4.	Additional works for the application of mineral materials (boreholes, injection)	-	-	-
5.	Cuboid concrete blocks	2500 pcs.	3.22 zł/piece	8 050.00 zł
6.	Performance of packwalls (1 running meter)		None	
7.	Other construction materials		None	
8.	Laboratory tests of gas samples for early detection of spontaneous fires conducted by the	3.538 pcs.	23 zł/piece	81 374.00 zł

	mining facility			
9.	Laboratory tests of gas samples for early detection of spontaneous fires conducted at an order	77 pcs.	215 zł/piece	16 555.00 zł
10.	Additional measurements (thermographic camera, pyrometer)	83	547.19	45 416.77 zł
11.	Use of inert gases – N_2 or CO_2			
12.	Inertization works		None	
13.	Lease or cost of use of inertization devices			
14.	Man-days related to preventing the fire hazard	450 man-days	362.33 zł/ man-day	163 048.50 zł
15.	Man-days of mine rescue workers working at preventive works	1 350 man-days	547.19 zł/ man-day	738 706.50 zł
16.	Use of water and other utilities	-	-	-
17.	**Total cost of the preventive works**			**1 118 397.77 zł**

Table 3. List of costs of climatic hazard prevention practices for longwall 1 [5, 7].

No.	Name of the cost	No./Amount	Unit cost (zł)	Total cost (zł)
1.	Air conditioning devices	8 pcs.	90 000.00 zł	720 000.00 zł
3.	Auxiliary fans	0 pcs.	-	-
4.	Fans with air-ducts	8 pcs.	70 000.00 zł	560 000.00 zł
5.	Operation of the devices	2 252 man-days	374.50 zł	843 374.00 zł
6.	Insulation of headings	-	-	-
7.	Auxiliary materials	-	-	-
8.	Intensification of ventilation	-	-	-
9.	Man-days related to preventing climatic hazard	563 man-days	379.36 zł/ man-day	213 579.68 zł
10.	Energy costs			
	Air conditioning devices x8	450 x 8 days	720.00 zł	2 592 000.00 zł
	Fans for the air coolers x8	450 days	154.08 zł	554 688.00 zł
11.	**Total cost of the preventive works**			**5 483 641.68 zł**

Table 4. List of costs of coal dust explosion prevention practices for longwall 1 [5, 7].

No.	Name of the cost	No./Amount	Unit cost (zł)	Total cost (zł)
1.	Rock dust	342 Mg	146.34 zł/Mg	50 048.28 zł
2.	Containers for water curtains	-	-	-
3.	Wooden/steel structures for explosion-proof barriers	Wood – 120 m^3 Hooks – 1920 pcs.	643.40 zł/m^3 2.15 zł/piece	77 208.00 zł 4 128.00 zł
4.	Monthly maintenance of protection zones against the propagation of the coal dust explosion in the region	18.5 months	17121 zł/month	316 738.50 zł
5.	Man-days related to the prevention of coal dust explosion hazard	3 378 man-days	379.36 zł/ man-day	1 281 478.08 zł
6.	Energy costs	-	-	-
7.	**Total cost of the preventive works**			**1 729 600.86 zł**

Table 5 presents the summary of aerological prevention costs borne during the exploitation of longwall 1.

Table 5. Summary of costs of methane prevention practices for longwall 1 [5, 7].

Cost of methane hazard prevention	19 570 966.02 zł
Cost of fire hazard prevention	1 118 397.77 zł
Cost of climatic hazard prevention	5 483 641.68 zł
Cost of dust hazard prevention	1 729 600.86 zł
Total cost of preventive ventilation practices in longwall No. 1	**27 902 606.33 zł**

3. Costs of prevention and cost indices

To assess the costs of aerological prevention, the following cost indices must be calculated:

- percentage of costs in proportion to the obtained revenue U_P, %,
- cost of prevention in conversion to 1 Mg of excavated coal K_{PMg}, zł/Mg.

Percentage of cost of prevention U_P was calculated based on the following formula:

$$U_P = \frac{K_P}{P_C} \cdot 100\%, (\%) \quad (1)$$

where:

K_P – total cost of preventive practices, zł,
P_c – total revenue, zł.

The total cost of the preventive works is the sum of the individual preventive ventilation practices costs.

The cost of prevention in conversion to 1 Mg of excavated coal KPMg was calculated based on the following formula:

$$K_{PMg} = \frac{K_P}{W_C}, \text{ (zł/Mg)} \quad (2)$$

where:

W_C – total output from the longwall, Mg.

The total cost of prevention in conversion to 1 Mg of the excavated coal K_{PMg} is the sum of the individual preventive practices costs in conversion to 1 Mg of the excavated coal.

To be able to compare the costs of the preventive works in relation to the coal price, a mean yearly price of 1 Mg of hard power coal was established. The price was calculated based on the coal price rate in line with the mean ARA prices [8] and – in the period in concern – it reached 70.42 USD. In line with the mean yearly exchange rate of USD in the years 2013-2015, which amounted to 3.46 zł [9], the price of 1 Mg of hard power coal was 243.65 zł/Mg.

18 longwalls exploited in the years 2013-2015 by one of the coal mining companies were subjected to the analysis [5]. The costs of the preventive practices and the costs as percentages have been exhibited in tables 6 and 7.

The highest costs of prevention of natural hazards related to ventilation are generated by preventive works linked to the limitation of the methane hazard. As shown in table 6, the lowest methane hazard prevention cost was 1 127 998.96 zł, while the highest reached as much as 23 080 212.22 zł. The span is thus considerable. Most of all, this is related to the range of the preventive works, especially the methane

IOP Conf. Series: Materials Science and Engineering **268** (2017) 012008 doi:10.1088/1757-899X/268/1/012008

removal from the rock mass. As it may be noted, no active methane prevention was conducted in case of 5 longwalls. The mean cost of the conducted methane prevention within the company was 5 528 116.48 zł.

Prevention against spontaneous fire hazard was conducted in case of all of the analyzed longwalls. The preventive works related to fire hazard are characterized by much lower costs, although longwalls exist in which the costs are also considerable. To exemplify this, the mean cost of preventive works in case of the analyzed longwalls was 2 677 879.64 zł. Only in case of 5 longwalls the costs of prevention exceeded the mean value. In case of two longwalls it was several-fold higher than the mean value and it amounted to 7 475 712.70 zł and 11 749 548.56 zł, respectively. The costs in case of the rest of the longwalls fell within the range of 167 963.73 zł – 4 858 262.01 zł. In that range, this resulted in a mean cost of prevention of 1 811 035.77 zł. It is also worth to note the individual costs borne within the fire prevention practices presented in the earlier example. The highest costs within these prevention works are the costs of the man-days. In the presented example, as much as 80.6% of the costs spent on prevention of fire hazard were spent on the man-days of the mining plant's rescue workers performing their duties in the form of preventive works and the man-days of the ventilation division's employees. The division of costs of fire prevention works in case of other analyzed longwalls is similar.

Preventive works related to climatic conditions were conducted only in 8 regions of the analysed longwalls. Their costs varied largely from 613 673.51 zł up to as much as 12 055 022.10 zł. The highest unit costs in case of prevention related to climatic conditions are denoted mostly by the costs of machines, air conditioning and auxiliary devices as well as the costs of the electric energy used to power the machines and devices.

From among all the costs borne for the preventive works conducted in relation to natural hazards in the range of ventilation, the lowest costs were related to the prevention of coal dust explosion hazard. It is one of the hazards with regard to which the prevention is obligatory [10]. The mean cost of the preventive works related to dust explosion hazard was 523 578.56 zł, although in case of two regions the costs amounted to 1 335 881.01 and 1 729 600.86 zł, respectively.

Figure 2 presents the costs of preventive works related to ventilation. As it may be noted, the highest cost of the conducted prevention is related to the methane hazard, while the lowest cost is related to the coal dust explosion hazard.

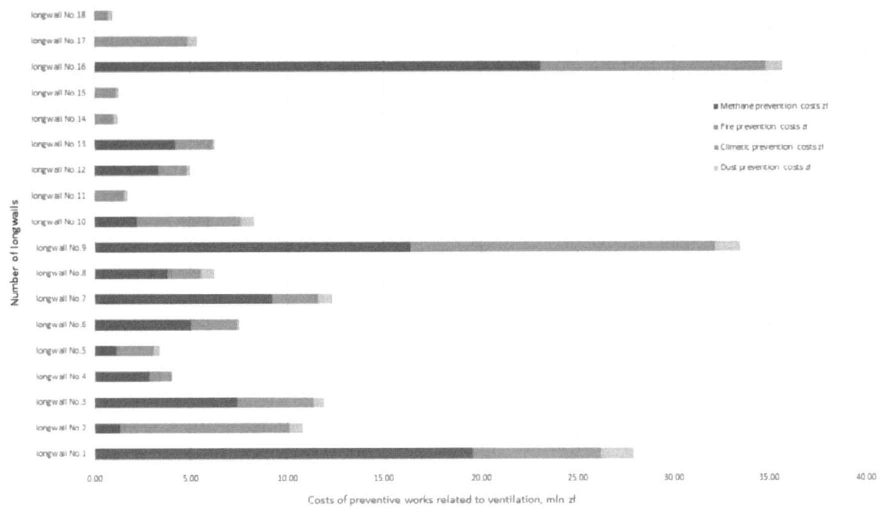

Figure 2. Costs of the aerological preventive practices in case of the analyzed longwalls.

IOP Conf. Series: Materials Science and Engineering **268** (2017) 012008 doi:10.1088/1757-899X/268/1/012008

Table 6. Costs of aerological preventive practices.

	Total output W_C (Mg)	Total revenue P_C (zł)	Methane prevention costs K_{PM} (zł)	Fire prevention costs K_{PP} (zł)	Climatic prevention costs K_{PK} (zł)	Dust prevention costs K_{PPL} (zł)	Total cost of the preventive works K_P (zł)
longwall No.1	965 532	235 251 871.80	19 570 966.02	1 118 397.77	5 483 641.68	1 729 600.86	27 902 606.33
longwall No.2	558 018	135 961 085.70	1 326 128.40	7 475 712.70	1 303 146.42	718 274.54	10 823 262.06
longwall No.3	1 303 819	317 675 499.35	7 388 397.34	1 840 149.90	2 120 459.40	582 193.60	11 931 200.24
longwall No.4	155 343	37 849 321.95	2 865 640.23	521 430.96	613 673.51	35 069.70	4 035 814.40
longwall No.5	682 836	166 372 991.40	1 127 998.96	167 963.73	1 808 183.04	287 664.80	3 391 810.53
longwall No.6	414 661	101 032 152.65	4 997 091.93	1 004 621.95	1 400 759.92	120 034.40	7 522 508.20
longwall No.7	389 392	94 875 360.80	9 212 301.90	2 399 711.70	Not used	750 630.02	12 362 643.62
longwall No.8	548 225	133 575 021.25	3 794 635.40	1 737 317.00	Not used	683 330.16	6 215 282.56
longwall No.9	561 807	136 884 275.55	16 377 089.90	3 694 356.77	12 055 022.10	1 335 881.01	33 462 349.78
longwall No.10	822 958	200 513 716.70	2 227 132.50	3 707 806.58	1 654 043.78	710 428.75	8 299 411.61
longwall No.11	346 417	84 404 502.05	Not used	1 573 923.75	Not used	177 868.78	1 751 792.53
longwall No.12	758 683	184 853 112.95	3 344 598.59	1 487 906.10	Not used	127 050.21	4 959 554.90
longwall No.13	578 793	141 022 914.45	4 193 903.32	1 957 206.60	Not used	127 050.21	6 278 160.13
longwall No.14	1 518 280	369 928 922.00	Not used	1 005 440.00	Not used	241 025.85	1 246 465.85
longwall No.15	333 550	81 269 457.50	Not used	1 151 312.45	Not used	149 835.78	1 301 148.23
longwall No.16	1 237 743	301 576 081.95	23 080 212.22	11 749 548.56	Not used	884 925.33	35 714 686.11
longwall No.17	487 795	118 851 251.75	Not used	4 858 262.01	Not used	526 839.26	5 385 101.27
longwall No.18	106 802	26 022 307.30	Not used	750 765.04	Not used	236 710.73	987 475.77
Mean	**653 925**	**159 328 880.39**	**5 528 116.48**	**2 677 879.64**	**1 468 829.44**	**523 578.56**	**10 198 404.12**
Total	**11 770 654**	**2 867 919 847.10**	**99 506 096.71**	**48 201 833.57**	**26 438 929.85**	**9 424 413.99**	**183 571 274.12**

IOP Conf. Series: Materials Science and Engineering **268** (2017) 012008 doi:10.1088/1757-899X/268/1/012008

Table 7. Cost indices of the aerological preventive practices.

	Cost percentage of methane prevention	Cost percentage of fire prevention	Cost percentage of climatic prevention	Cost percentage of dust prevention	Total percentage of the prevention costs	Cost of methane prevention per 1 Mg of the output	Cost of fire prevention per 1 Mg of the output	Cost of climatic prevention per 1 Mg of the output	Cost of dust prevention per 1 Mg of the output	Total cost of prevention per 1 Mg of output
	U_{PMe} (%)	U_{PPo} (%)	U_{PKl} (%)	U_{PPy} (%)	U_P (%)	K_{PMgMe} (zł)	K_{PMgPo} (zł)	K_{PMgKl} (zł)	K_{PMgPy} (zł)	K_{PMg} (zł)
longwall No.1	8.32	0.48	2.33	0.74	11.86	20.27	1.16	5.68	1.79	28.90
longwall No.2	0.98	5.50	0.96	0.53	7.96	2.38	13.40	2.34	1.29	19.40
longwall No.3	2.33	0.58	0.67	0.18	3.76	5.67	1.41	1.63	0.45	9.15
longwall No.4	7.57	1.38	1.62	0.09	10.66	18.45	3.36	3.95	0.23	25.98
longwall No.5	0.68	0.10	1.09	0.17	2.04	1.65	0.25	2.65	0.42	4.97
longwall No.6	4.95	0.99	1.39	0.12	7.45	12.05	2.42	3.38	0.29	18.14
longwall No.7	9.71	2.53	0.00	0.79	13.03	23.66	6.16	0.00	1.93	31.75
longwall No.8	2.84	1.30	0.00	0.51	4.65	6.92	3.17	0.00	1.25	11.34
longwall No.9	11.96	2.70	8.81	0.98	24.45	29.15	6.58	21.46	2.38	59.56
longwall No.10	1.11	1.85	0.82	0.35	4.14	2.71	4.51	2.01	0.86	10.08
longwall No.11	0.00	1.86	0.00	0.21	2.08	0.00	4.54	0.00	0.51	5.06
longwall No.12	1.81	0.80	0.00	0.07	2.68	4.41	1.96	0.00	0.17	6.54
longwall No.13	2.97	1.39	0.00	0.09	4.45	7.25	3.38	0.00	0.22	10.85
longwall No.14	0.00	0.27	0.00	0.07	0.34	0.00	0.66	0.00	0.16	0.82
longwall No.15	0.00	1.42	0.00	0.18	1.60	0.00	3.45	0.00	0.45	3.90
longwall No.16	7.65	3.90	0.00	0.29	11.84	18.65	9.49	0.00	0.71	28.85
longwall No.17	0.00	4.09	0.00	0.44	4.53	0.00	9.96	0.00	1.08	11.04
longwall No.18	0.00	2.89	0.00	0.91	3.79	0.00	7.03	0.00	2.22	9.25
average	**3.49**	**1.89**	**0.98**	**0.37**	**6.74**	**8.51**	**4.60**	**2.39**	**0.91**	**16.42**

The total cost of the conducted preventive works is relatively variable. Depending on the occurring natural hazards and the necessary scope of preventive works, the cost may be considerable. This may be observed while comparing the costs of the preventive works with the total revenue obtained from the longwalls (figure 3).

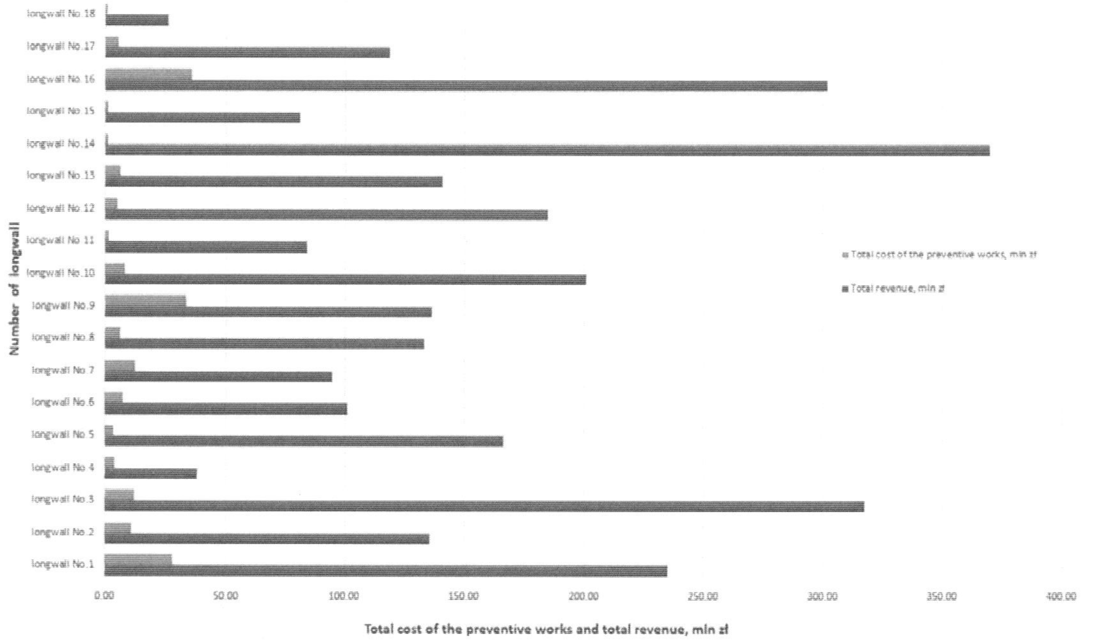

Figure 3. A comparison of the total cost of the aerological preventive practices and the total revenue in case of the analyzed longwalls.

One of the cost indicators for a given prevention practice is its share in the total revenue. While analyzing the shares of the costs (table 7, figure 4) the highest costs concern the methane hazard prevention, amounting to a mean percentage of 3.49%, although longwalls exist for which the value varies around 10%.

In case of the fire hazard, the mean share of the costs reached 1.89%. The span of the cost fractions was wide and varied in the range of 0.10 to 5.50%. In case of the thermal hazard, the percentage of the prevention costs varied in the range of 0.67-8.81%. The lowest portion of the costs was exhibited in case of coal dust explosion hazard prevention. The mean percentage was 0.37%, while the portion of the costs in the total revenue varied between 0.07 to 0.98%.

The cost of a given prevention practice per 1 Mg of output is a significant indicator (table 7, figure 5). The mean cost of methane hazard prevention was 8.51 zł per a tonne of output. The lowest cost thereof was 1.65 zł/1 Mg while the highest reached 29.15 zł/1 Mg. In case of the fire hazard prevention, the mean cost was 4.60 zł/1 Mg, while the costs of prevention per 1 Mg of output varied in the range of 0.25-13.40 zł/1 Mg. The necessity to exercise the prevention of thermal hazard requires high costs to be borne. In case of the analyzed longwalls, the range of the costs of the thermal prevention per 1 Mg of output varied in the range from 1.63 zł/1 Mg up to 21.46 zł/1 Mg. The prevention against coal dust explosion was the most "even" in terms of the costs. Its mean cost was 0.91 zł/1 Mg in the range of 0.16-2.38 zł/1 Mg.

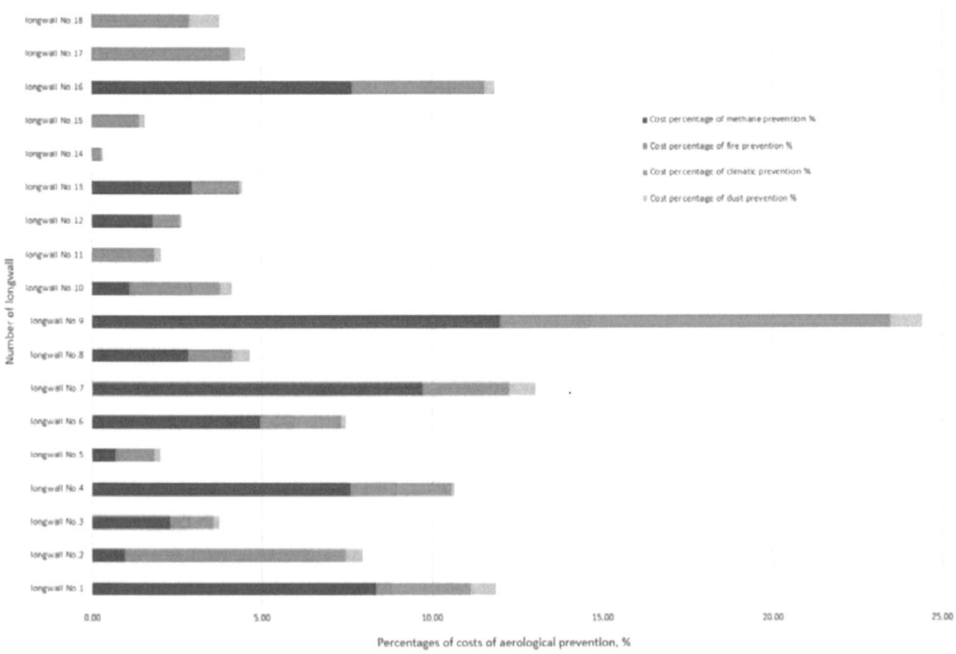

Figure 4. Percentages of costs of aerological prevention practices in the total revenue in case of the analyzed longwalls.

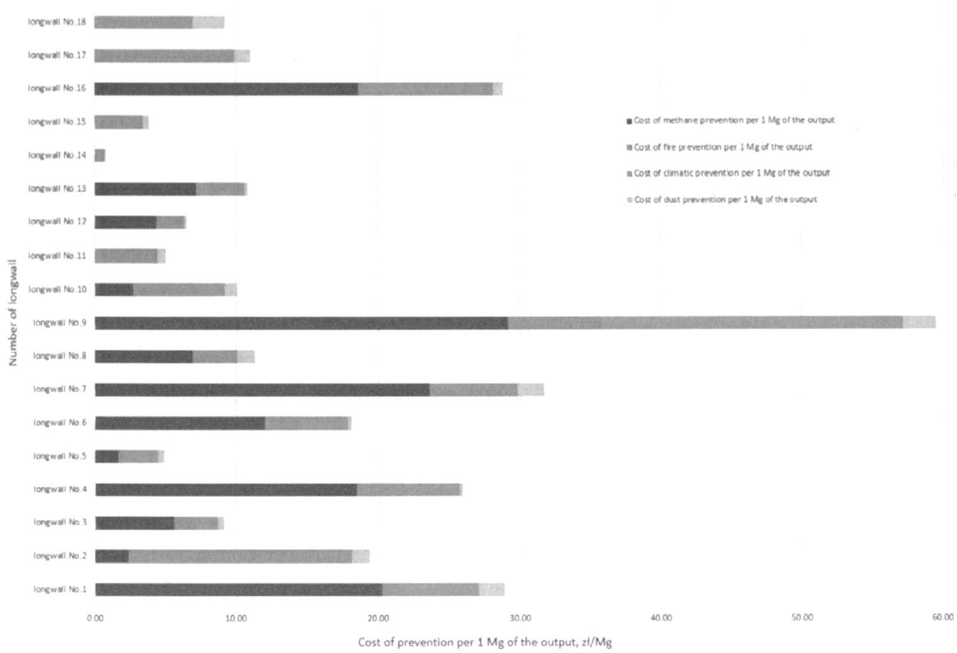

Figure 5. Costs of aerological prevention practices per 1 Mg of output for the analyzed longwalls.

4. Summary

The performed analysis of costs of preventive works exhibits that the highest costs are borne due to fighting the methane hazard. The large majority of the analyzed longwalls, namely 72.2% required the application of methane hazard prevention. The costs of methane prevention constitute as much as 51.8% of the total costs borne by the coal mines for the prevention of aerological hazards. The mean cost of methane hazard prevention amounted to 8.51 zł/1 Mg of output. One may note, however, that often the costs of methane prevention exceed 20.00 zł/1 Mg (maximum of 29.15 zł/1 Mg).

The prevention works that were most often conducted were the works in the scope of spontaneous fire prevention. In case of all the analyzed longwalls, active prevention of fire hazard was conducted. The mean cost of the fire prevention was 4.60 zł/1 Mg.

In case of coal dust explosion hazard prevention, due to its obligatory character, the costs of application are lower as compared to other prevention practices. This is related to the to the legally required algorithm of conduct imposed by means of the mining provisions. The mean cost of prevention amounted to 0.91 zł/1 Mg of output.

Thermal hazard prevention, although costly, was not often applied in case of the analyzed longwalls. Active prevention of thermal hazard was conducted only in 8 of the 18 analyzed longwalls.

The total participation of the costs of preventive works in the revenue of the analyzed longwalls was 6.74%, while the total mean cost of prevention practices was 16.42 zł/1 Mg of output.

The share of the prevention practices costs related to ventilation increases gradually. This is not only related to the significant decrease in the price of coal, but also the necessity to exploit seams at a gradual and slow increase of the impact of natural hazards and the increasingly difficult geological and mining conditions. If the low prices are maintained at a stable level for a longer period, a situation may occur when the costs of preventing the hazards related to ventilation shall be so high that the exploitation of coal will prove unprofitable.

References

[1] Gawlik L 2005 Koszty stałe i zmienne w kopalniach węgla kamiennego, *Polityka energetyczna*, **8** pp 331–45

[2] Gawlik L 2007 Koszty stałe i zmienne pozyskania węgla kamiennego jako element zarządzania, *Polityka energetyczna*, **10** pp 471–82

[3] Musioł D 2016 Działania w zakresie profilaktyk aerologicznych jako niezbędny element nakładów na wydobycie w ścianach eksploatacyjnych, *Systemy wspomagania w inżynierii produkcji, Górnictwo – perspektywy i zagrożenia,* **1 (13)** pp 122–37

[4] Sułkowski J and Musioł D 2008 Effect of bed splitting on fighting aerologic hazards in exploitation sections of hard coal mines *Archives of Mining Sciences* **53**/4 pp 545–54

[5] Documentation of preventive works for longwalls in a coal mining company 2016 – not published

[6] Technical designs for longwalls in a coal mining company, 2016 – not published

[7] Costs of preventive works for longwalls in a coal mining company 2016 – not published

[8] Coal price rates by ARA (Amsterdam-Rotterdam Antwerpia) - http://gornictwo.wnp.pl

[9] Exchange rates: http://www.bankier.pl/waluty/kursy-walut/nbp/USD

[10] Rozporządzenie Ministra Gospodarki z dnia 28 czerwca 2002 r. (z późniejszymi zmianami) w sprawie bezpieczeństwa i higieny pracy, prowadzenia ruchu oraz specjalistycznego zabezpieczenia przeciwpożarowego w podziemnych zakładach górniczych, Dz. U. Nr 139, poz. 1169

IOP Conf. Series: Materials Science and Engineering **268** (2017) 012009 doi:10.1088/1757-899X/268/1/012009

Possibility of reducing costs of mining operations – economic aspects of workplace accidents

Adam Duda

Faculty of Mining and Geology, Silesian University of Technology, Akademicka 2, 44-100 Gliwice, Poland.

adam.duda@polsl.pl

Abstract. The article presents methods of calculating costs of workplace accidents incurred by an employer, and the influence of the number and severity of accidents on changes in the amount of accident insurance contribution paid by an employer within the framework of the social security system.

1. Introduction

Work accidents are still a significant problem for employers, and, first of all, for the society. In 2016, in Poland, 87 183 people were injured in accidents, including 464 severe injuries and 239 fatalities. The factors which determine solutions chosen by employers, also in work-related safety, are usually economic ones. That is why it is so important to know how accidents impact the economic performance of an enterprise.

The issue of analysing costs of work accidents is nothing new, and the United States of America is the country which has paid a lot of attention to the problem. The reason may be the fact that the issue of accident costs is particularly important for employers in the countries where the burden of possible consequences of an accident, to a large extent, falls on an employer, rather than the society. The first ones to research the issue were H.W Heinrich [1], R.M. Simonds, and J.V. Grimaldi [2]. In Poland, in the interwar years, research on the costs of accidents was conducted by W. Adamiecki, A. Anasiewicz and E. Modliński. Even after the years, the main assumptions in calculating the costs have not changed. Works by H.W. Heinrich, R.M. Simonds, P. Compes and A. Soderqvist remain the foundation of the methods currently in use [3]. The methods are modified in the area of identifying direct and indirect costs and linking them with the benefits an enterprise draws from the high level of work related safety [4]. Nowadays, methods of calculating accidents costs are developed by e.g. M. Oxenburgh, P. Marlow, C. Zangemeister, U. Johanson and Marc de Greef. In Poland works by J. Rzepecki and Z. Pawłowska have a big influence on the research on accident costs and methods of calculating them.

2. Methods of calculating costs of work accidents in Poland

Nowadays in Poland the most common is calculating costs of accidents following J. Rzepecki's proposal applied by the Central Institute for Labour Protection. According to its guidelines cost of accident is [5, 6]:

$$K_{SO} = K_{ZO} + K_{PO} + K_{PSO} \tag{1}$$

IOP Conf. Series: Materials Science and Engineering **268** (2017) 012009 doi:10.1088/1757-899X/268/1/012009

where:
K_{SO} – total social cost of a work accident,
K_{ZO} – total cost incurred by the enterprise,
K_{PO} – total cost incurred by the victim and their family,
K_{PSO} – total cost incurred by the society.

Costs incurred by the society (K_{PSO}) are costs covered by the Social Insurance Institution (ZUS) or costs of treatment covered by the National Health Fund. They are e.g.:

- sickness benefit,
- rehabilitation benefit,
- complementary benefit,
- single indemnity,
- incapacity benefit, including training benefit,
- family allowance pension, additional payment to family allowance pension,
- nursing supplement,
- additional payment to family allowance pension for a double orphan,
- covering costs of dental treatment and vaccinations and providing orthopaedic items and aids specified by the Act.

The costs incurred by an enterprise include basic cost components which can be calculated with the following equation [6]:

$$K_{ZO} = (K_{CS} + K_{PMiT} + K_N + K_Z + K_{ZP} + K_{SM} + K_{NP} + K_{OW} + K_I) - O \quad (2)$$

where:
K_O – total cost,
K_{CS} – cost of lost time,
K_{PMiT} – cost of first aid and transport,
K_N – cost of overtime,
K_Z – cost of replacement personnel,
K_{ZP} – costs of interruptions in production,
K_{SM} – cost of material losses,
K_{NP} – cost of repairs,
K_{OW} – cost of financial compesation
O – compensation received from insurance institutions.

As the above equation shows, calculating costs of accidents which are incurred by an employer is not difficult. The problem is availability of data. The accounting system used by employers makes it hard to obtain data on to what extent the costs associated with accidents can be directly extracted basing on analytical accounts. Lack of easy identification of accident costs in the accounting system makes it hard, yet not impossible, to analyse costs of accidents. It is possible to perform such an analysis when data, even raw ones, are available for an analyst. Such analyses are conducted in many enterprises, particularly in the ones which, due to an accident, receive compensations from insuring companies. However, the data are rarely shared with third parties, which is fully understandable, but results in lack of such data for research.

Data concerning costs of accidents incurred in one of coal mines in 2006-2008 showed that the average cost of an accident incurred by the employer was then approximately 8 000.00 PLN. What is important, the analysed accidents did not stop the production. They were work accidents where there was no material damage and the employer did not incur any costs of inter-rupted production process. The accidents which interrupt (lower) production, also cause huge losses – most often they are ca-tastrophes or incidents caused by forces of nature of great potential, in which employees are victims.

In the literature, it is assumed that the cost of an accident incurred by an employer is 25% of the total cost of an accident i.e. approximately 75% of the costs is incurred by 'the society' namely the state budget (ZUS, NFZ)[3].

Such a large share of the costs of accidents incurred by the society means that the State tries, through various mechanisms, 11.00 promote safety at work and 'reward' the employers with the fewer and less severe accidents than other employers within the same business activity (PKD).

3. Differentiating accident insurance contribution depending on the risk category

By implementing motivational social security systems, it is possible to make employers perceive work related safety as an economic benefit. In Poland, regulations which enable differentiating the amount of accident insurance contribution were introduced by the regulation of the Minister of Labour and Social Policy of 29 November 2002 on differentiating social insurance contribution in respect of accidents at work and occupational diseases in relation to occupational hazards and their effects (Journal of Laws of 2002, No. 200, item 1692 as amended). According to the regulation, the amount of accident insurance contribution for a business activity is calculated on the basis of the risk category for a business activity determined basing on partial risk categories K_1, K_2, K_3, K_4 following the equation below [7]:

$$K_{DX} = \frac{K_1 + K_2 + K_3 + K_4}{4} \quad (3)$$

where:

K_{DX} – risk category for given business activity,

K_1 – risk category corresponding the rate of injuries in work accidents, total,

K_2 – risk category corresponding the rate of fatalities and severe injuries in work accidents,

K_3 – risk category corresponding the rate of personnel working in hazardous conditions, occurring when the highest admissible concentrations and intensity of factors detrimental to health in the work environment are exceeded,

K_4 – risk category corresponding the rate of occupational diseases.

To determine given risk categories, it is necessary to start with calculating the rate of:

- total injuries in work accidents,
- fatalities and severe injuries in work accidents,
- personnel working in hazardous conditions, occurring when the highest admissible concentrations and intensity of factors detrimental to health in the work environment are exceeded,
- recorded occupational diseases.

Given indices are calculated per 1 000 employees and determined separately for a full calendar year for business activities. Basing on the dependences, risk category for a business activity is calculated and together with accident insurance contribution they are attached to the aforementioned regulation.

Another element which, together with risk categories, influences the interest rate of an accident insurance contribution deter-mined for given contribution payers within their business activities, i.e. its increase or decrease, is so-called adjustment factor. The adjustment factor is determined individually for each contribution payer i.e. basing on the difference between risk cate-gory specified in the regulation for a business activity, which given contribution payer belong to, and given contribution payer's risk category determined with the dependence [7]:

$$K_{ZX} = \frac{K_1 + K_2 + 2K_3}{4} \quad (4)$$

where:

K_{ZX} – risk category for each given contribution payer.

The basis to determine partial risk categories are arithmetic means of the rates of [7]:

- total injuries in work accidents (W_1),
- fatalities and severe injuries in work accidents (W_2),
- personnel working in hazardous conditions, occurring when the highest admissible concentrations and intensity of factors detrimental to health in the work environment are exceeded (W_3).

The rates are calculated per 1 000 insured persons and determined for contribution payers separately for a full calendar year following the equations [7]:

$$W_1 = \frac{P_1}{U} \times 1000 \tag{5}$$

$$W_2 = \frac{P_2}{U} \times 1000 \tag{6}$$

$$W_3 = \frac{P_3}{U} \times 1000 \tag{7}$$

where:
P_1 – total number of victims injured in work accidents;
P_2 – number of victims injured in fatal and severe work accidents;
P_3 – number of personnel working in hazardous conditions, occurring when the highest admissible concentrations and intensity of factors detrimental to health in the work environment are exceeded;
U – number of insured persons,

Arithmetic means of rates W_1, W_2, W_3 are the basis to determine partial risk categories K_1, K_2, K_3. Arithmetic means of the rates are calculated basing on the last three consecutive calendar years.

As the above dependences show, the number of victims injured in work accidents and their severity determine which risk cate-gory a contribution payer i.e. employer will fall into.

After calculating risk category for an employer it has to be compared with risk category for a business activity (PKD) and apply adjustment factor of [7]:
0.5 - if the risk category determined for contribution payer is by at least 6 categories lower than the risk category determined for a business activity;
0.6 - if the risk category determined for contribution payer is by 5 categories lower than the risk category determined for a business activity;
0.7 - if the risk category determined for contribution payer is by 4 categories lower than the risk category determined for a business activity;
0.8 - if the risk category determined for contribution payer is by 3 categories lower than the risk category determined for a business activity;
0.9 - if the risk category determined for contribution payer is by 2 categories lower than the risk category determined for a business activity;
1.1 - if the risk category determined for contribution payer is by 2 categories higher than the risk category determined for a business activity;
1.2 - if the risk category determined for contribution payer is by 3 categories higher than the risk category determined for a business activity;
1.3 - if the risk category determined for contribution payer is by 4 categories higher than the risk category determined for a business activity;
1.4 - if the risk category determined for contribution payer is by 5 categories higher than the risk category determined for a business activity;
1.5 - if the risk category determined for contribution payer is by at least 6 categories higher than the risk category determined for a business activity;
1.0 – in other cases.

According to current legal regulations, in extreme cases, an employer will pay either 50% more or 50% less for accident insurance. Business activity: 'Mining of hard coal and lignite' falls into risk category 12 and the interest rate of the contribution is 3.33% [7]. The number may not reflect the financial scale obtained due to differentiating accident insurance contribution hence it is worth to simulate it. Assumptions and results of such a simulation are presented in table 1.

Table 1. Simulated changes in accident insurance contribution.

	Year							
Data	2014	2015	2016	2017	2018	2019	2020	2021
Number of people injured	3382	3370	3310	3254	3198	3121	3100	3091
Number of people injured in accidents	47	38	50	41	21	18	23	
Number of fatalities and persons severely injured in accidents	1	0	0	0	0	0	0	
Number of personnel working in dangerous conditions	983	971	951	942	921	901	894	
Rate of total injuries (W_1)	13.897	11.276	15.106	12.600	6.567	5.767	7.419	
Rate of fatalities and severe injuries (W_2)	0.296	0.000	0.000	0.000	0.000	0.000	0.000	
Rate of personnel working in dangerous conditions (W_3)	290.66	288.13	287.31	289.49	287.99	288.69	288.39	
Rate calculated on the basis of the last three calendar years to determine partial categories (W_1, W_2, W_3)								
Rate of total injuries (W_1)				13.4	13.0	11.4	8.3	6.6
Rate of fatalities and severe injuries (W_2)				0.099	0.000	0.000	0.000	0.000
Rate of personnel working in dangerous conditions (W_3)				289	288	288	289	288
Risk category K_1				10	10	9	7	6
Risk category K_2				5	1	1	1	1
Risk category K_3				9	9	9	9	9
Risk category for given business activity (K_{DX})				12	12	12	12	12
Contribution for given business activity				3.33%	3.33%	3.33%	3.33%	3.33%
Risk category for given employer (K_{ZX})				8	7	7	7	6
Difference between categories				4	5	5	6	6
Adjustment factor				0.7	0.6	0.6	0.5	0.5

Contribution calculated for given employer	2.33%	2.00%	2.00%	1.67%	1.67%
Average monthly remuneration for calculating accident insurance contribution (PLN)	6000	6000	6000	6000	6000
Total contributions paid by employer in one year (PLN)	5461253	4600515	4489746	3716280	3705491

For the presented calculations it was assumed that:

- in consecutive years number of accidents will halve,
- value of a risk category and contribution for a business activity is constant,
- number of employees working in conditions of exposure in relation to the number of insured persons will not change significantly (with such an assumption it is possible to presents benefits of decreasing the number of accidents)
- remuneration, which the contribution is calculated upon, is constant

The changes in the amount of accident insurance contribution, which an employer will have to pay in consecutive years are presented in figure 1.

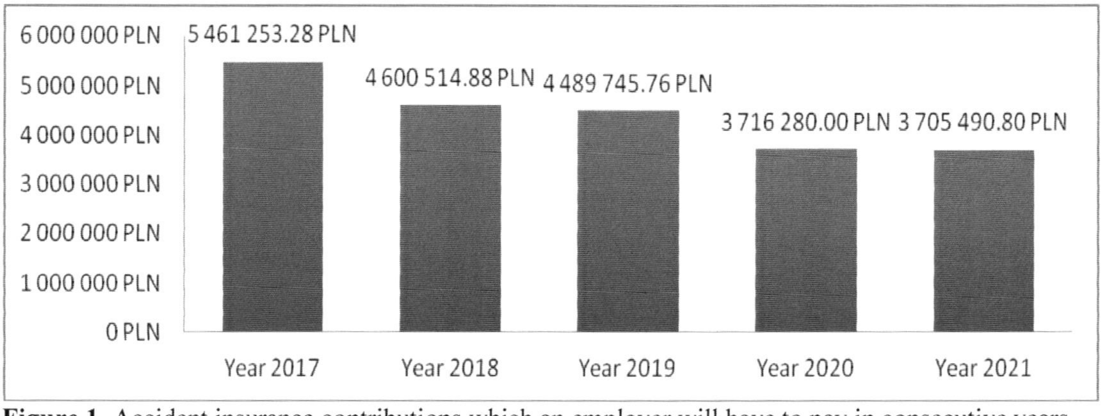

Figure 1. Accident insurance contributions which an employer will have to pay in consecutive years.

As the above graph shows, an employer, in relation to 2017, will pay a contribution lower: in 2018 by 860 738.40 PLN; in 2019 by 971 507.52 PLN; in 2020 by 1 744 973.28 PLN; in 2010 by 1 755 762.48 PLN. It has to be emphasised that the contribution was lowered from the value of accident insurance contribution already decreased by adjustment factor of 0.7.

4. Conclusions

It is an employer who has the greatest influence on creating safe working conditions. Costs of accidents incurred by an employer are not high enough to motivate them to take actions aimed at improving work-related safety – especially in case of the accidents which do not interrupt production or cause material damage, i.e. personnel's most common accidents.

Nearly 75% of costs of accidents is incurred by the society and the victims themselves. Costs incurred by the society are paid with money collected by the Social Insurance Institution which come from accident insurance contributions and the financial means of the National Health Fund.

One of the methods to motivate employers to improve work-related safety in an enterprise is differentiating accident insurance contributions.

IOP Conf. Series: Materials Science and Engineering **268** (2017) 012009 doi:10.1088/1757-899X/268/1/012009

Results of the simulated change in the amount of accident insurance contribution paid by an employer depending on the number of accidents show that reducing the number by approximately 50% an employer may lower, within 4 years, accident insurance contribution by 5 332 981.68 PLN.

The mechanism of differentiating accident insurance contribution must not be 'skewed' by employers' actions aimed at 'moving' accidents outside the workplace (accidents on the way to and from work), or creating systems of bonuses 'motivating' employees not to report accidents.

References

[1] Heinrich H W 1941 *Industrial Accident Prevention* (New York: A Scientific Approach McGraw-Hill)

[2] Simonds R M and Grimaldi J V 1956 *Safety management: Accident cost and control* (Illinois: Homewood Richard D. Irvin)

[3] Duda A 2010 Koszty wypadków przy pracy ponoszenie przez pracodawcę na przykładzie analizy kosztów w latach 2006-2008 w KWK "X" (Gliwice: Górnictwo zrównoważonego rozwoju Wydział Górnictwa i Geologii Politechniki Śląskiej CD-ROM) pp 1-6

[4] Szlązak J and Grodzicka A 2016 Ocena kosztów wypadków przy pracy i chorób zawodowych z punktu widzenia państw Unii Europejskiej *Wiadomości Górnicze* **4** pp 262–70.

[5] Rzepecki J 2005 Społeczne koszty wypadków przy pracy w Polsce. *Bezpieczeństwo Pracy: Nauka i Praktyka*, **7-8** pp 34-7.

[6] Rzepecki J 2015 Społeczne koszty wypadków przy pracy. *Bezpieczeństwo Pracy: Nauka i Praktyka* **5** pp 20-3.

[7] Rozporządzenie Ministra Pracy i Polityki Społecznej z dnia 29 listopada 2002 r. w sprawie różnicowania stopy procentowej składki na ubezpieczenie społeczne z tytułu wypadków przy pracy i chorób zawodowych w zależności od zagrożeń zawodowych i ich skutków (Dz. U. 2002, Nr 200, poz. 1692 ze zm.)

IOP Conf. Series: Materials Science and Engineering **268** (2017) 012010 doi:10.1088/1757-899X/268/1/012010

Mining machines effectiveness and OEE Indicator

Jacek Korski [1], Katarzyna Tobór-Osadnik[2] and Małgorzata Wyganowska[2]

[1]Famur SA, Katowice, Poland

[2]Silesian University of Technology, Gliwice, Poland

malgorzata.wyganowska@polsl.pl

Abstract. The situation in the hard coal industry in Poland is forcing the identification of effectual and practical indicators of the effectiveness of machinery and equipment. In the article, the authors discuss the possible use of the OEE indicator for the evaluation of production processes in hard-coal mines. In summary, recommendations are made to enable efficiency assessment of mining machinery using the OEE.

1. Introduction

Eliyahu Goldratt [1] noted that good methods and tools in the management of business and manufacturing used in several Japanese car factories proved to be ineffective in others. The main cause of these setbacks was found to be the difference in the processes and technologies employed by the latter, in comparison to the processes performed in the automotive industry.

Effectiveness in the employment of available resources in industrial processes, including human labor and technical equipment, in order to achieve efficiency is one of the basic management objectives, particularly with regards to operational management.

Mining processes have for many years been treated as production processes and attempts have been made for an equally long time, to implement methods and management tools transferred from other types of business, not just industrial ones [2]. Professor Bolesław Krupiński, a prominent Polish mining expert, paid close attention to the fact that mining processes are of a different nature, namely: "... the mine is, first and foremost, a great transport undertaking ...". Disparity between the processes used in extraction from those in production is one of the reasons for the failure of management methods derived from the Japanese car industry, when implemented in mining [3]. Still, hard coal mines are attempting to implement tools and methods, including diagnostic tools and methods, which will not only be ultimately useless but will also drain unnecessary resources from their implementation. The reason for this lies in the differences between both process systems.

One such attempt is in the employment of the Overall Equipment Effectivness (OEE) tool for evaluation of machine efficiency [4]. This indicator is an effective tool for evaluating the use of machines in production plants and is geared towards the diagnosis of machinery utilisation in the manufacturing process, i.e. at the part of the process system that is adapt towards the external customer, creating added value for the company. In production processes, especially in manufacturing, the high degree of utilisation of each machine in the area of primary production is one of the conditions of process efficiency [5]. In underground mines the basic process is the mining process, which involves extraction and transport processes related to the transport of minerals [6]. Both extraction and transport processes may be mechanized in different ways. Technical solutions and the organisation of work used in the extraction of minerals such as hard coal and the manner in which it is mechanised significantly affect the degree to which individual machines are utilised. It is important to note that the use of different evaluation criteria (indicators) is indicated by the theoretical and practical experiences of large mining companies in the world [7, 8].

2. OEE indicator

The Overall Equipment Effectiveness (OEE) rating is one of the Key Performance Indicators (KPIs) used in the TPM (Total Production Maintenance) system. This is an indicator of the equipment effectiveness used in the area of operational production management. Pioneering the implementation of TPM was the Japanese company Denso, which is part the Toyota Group, thus belonging to the automotive industry. The TPM system, being an element of Lean Management, consists of the preventative maintenance of machinery and equipment within the company by operators (autonomous maintenance) and maintenance personnel (scheduled maintenance). Adequate cooperation between maintenance and production allows for an improvement in the efficiency of even heavily used machinery and a significant reduction in hazards such as machinery breakdown, or unplanned downtime. Consequently, this serves to improve the efficiency of the manner in which machines and equipment are utilised.

OEE is the product of 3 other simpler indicators:

$$\textbf{OEE} = \textbf{Availability} * \textbf{Performance} * \textbf{Quality} \quad (1)$$

where:

Availability - The ratio of the time planned for a production task to the time that can actually be devoted to this task. Availability is reduced by machine down time stemming from breakdowns, retooling and the setting up of machinery (and according to the adopted method), as well as by the loss of speed of operation.

Performance – the ratio of availability to operating time.

Quality – the ratio of good and defective units produced to all units produced.

Figure 1 shows the OEE indicator in pictorial form.

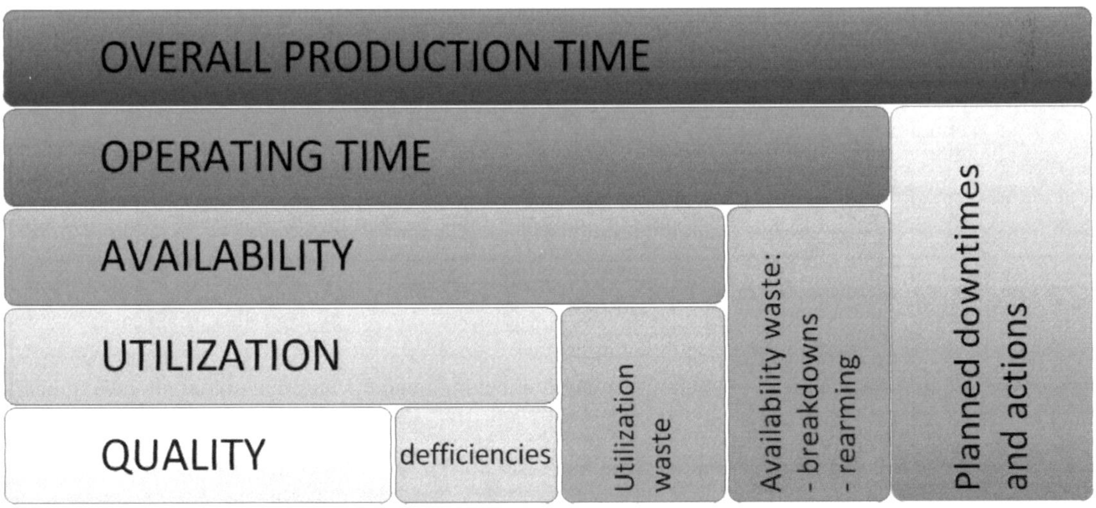

Figure 1. OEE structure (Source: according to [5])

The OEE indicator was developed to assess the efficiency of equipment in manufacturing processes where machine parts are fixed and unmovable, and only materials, semi-finished and finished goods move. Doubts regarding the availability parameter of the equipment in terms of the retooling and setting of the machines are due to the fact that the OEE was originally used in assembly-line processes. Despite being composed of machinery and equipment, the use of the OEE indicator encounters significant problems in unit production.

It should be noted that the production process most often takes place in predetermined physical conditions (eg. in buildings isolated from weather conditions), so the impact of typically natural conditions is negligible. A well-designed production system should work continuously during its operating time, and the drop in availability should result solely from breakdowns. Falls in equipment utilization are mainly a consequence of:

- „bottle-necks" (restrictions),
- failures in the provision of raw materials,
- other organisational failures.

Owing to the characteristics of this indicator, the question arises as to whether the OEE indicator be used effectively to assess the efficiency of machinery and equipment in hard coal mining.

3. Process systems in underground mining

Durlik [9] recognizes the basic process in a mine to be extraction, therefore making it a production process. However, using a commonly used classification of processes in an enterprise, with divisions into primary processes, ancillary processes and management processes [10], it should be noted that in a mine, the primary process is not a production process in this sense. The extraction process does not create the usefulness of the shape (character) of the material, which is typical for a production process.

For the external customer (the receiver of the process result), the extraction process creates the usability of space and time that is typical for logistic processes [6]. In the extraction process, sub-processes are not involved. These, however, form a key component in production processes. The share of operations of a technological nature (change of shape or physical or chemical properties) in the entire system of processes in a mine, i.e. complex structure of activities, is small.

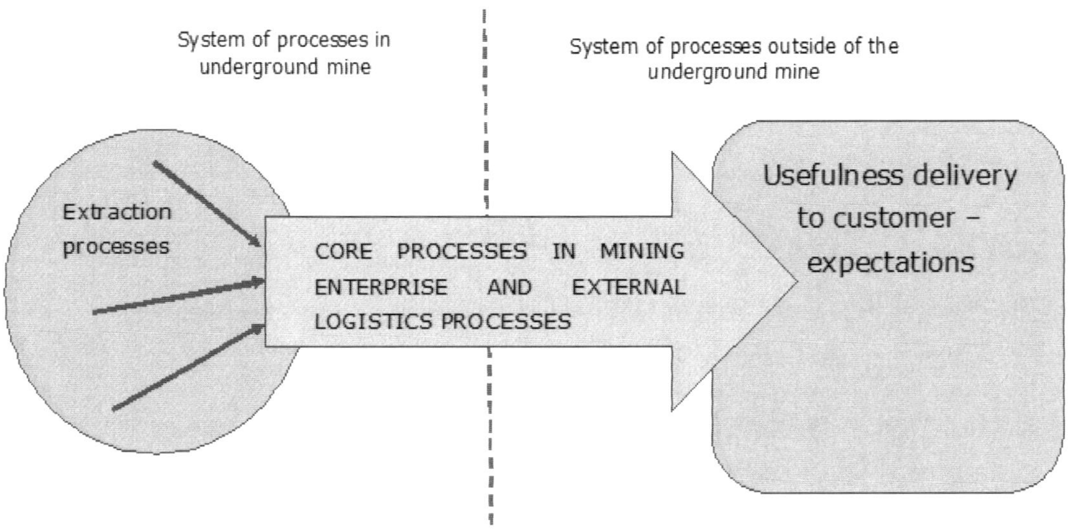

Figure 2. The primary process overall – the extraction process in a mine (based on [6]).

The basic process in the mining sector, ie the mining process, includes the sub-processes of excavation and transport of the extration material (figure 2). It should be noted that the transport of waste can be performed in a continuous system (eg conveyor belt, etc.), intermittent (vehicles, rail or shaft lift) or mixed. Particularly in terms of intermittent transport, there is usually a return drive element (unladen) that is difficult to define in the performance index of OEE equipment. It is difficult to indicate a uniform quality meter when transporting waste. By concentrating on the mining aspect of the core process of a mine, ie the mining process (excavation works + transport of the output), the mining process can be defined as: 'Emptying the mineral store previously accumulated by means of natural processes and supplying this mineral to the end user (external customer)' [6].

The mining process is only one of the components of a complex network of mine processes, as shown in figure 3. In this figure, the process of refining / enrichment of the mineral has been omitted as it is not always performed in a mine.

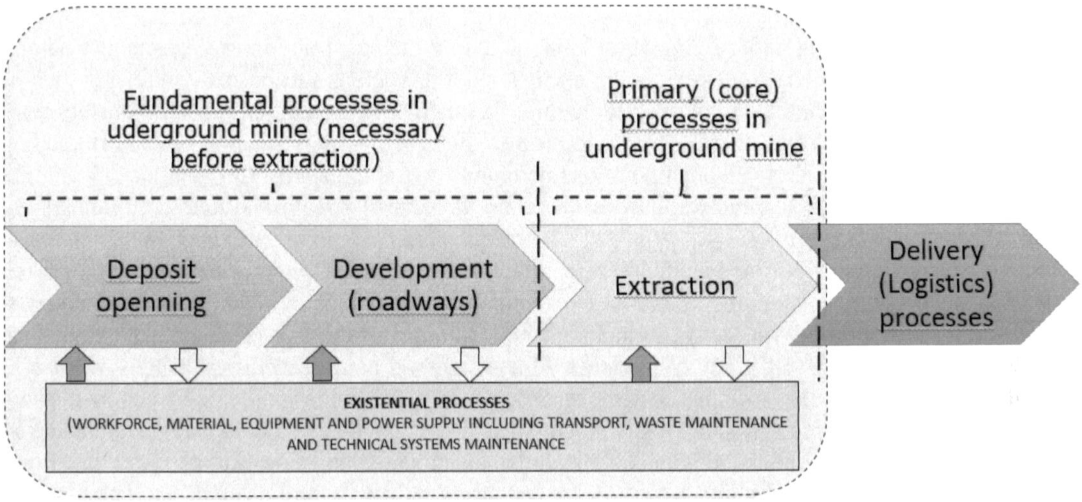

Figure 3. A generalised model of the mining process network.

Depending on the local conditions and constraints, as well as legal regulations and economic opportunities, different excavation systems are used in underground mining. Despite the variety of excavation processes, the basic cycle is shown in figure 4.

Figure 4. Basic continuous cycle of activities in the excavation process.

Differences in the selection of mining systems in underground mines also depend on the technology of individual actions and the degree to which they are mechanised. As an example, it is possible to say that today, excavation by hand also involves the use of mechanical tools, explosive materials (blast-hole drilling with manual tools, drilling machinery and, the rarely used, hydraulic excavation). Some mining methods impose a serial organisation of work (eg in the case of explosive materials), which, in the case of machinery and specialised equipment performing a single operation, results in a low, although reasonably justified, degree of exploitation which deteriorates one of the elements of the OEE.

IOP Conf. Series: Materials Science and Engineering **268** (2017) 012010 doi:10.1088/1757-899X/268/1/012010

4. Basic mechanisation systems in mining and the extent to which machinery and equipment is used

As mentioned in the previous section, traditional mining methods are still used in underground mining (by hand with only perhaps mechanised tools), partly mechanised with certain operations or actions being mechanised, and fully-mechanised systems.

In traditional and partially mechanised systems, operations are performed in series - sequentially, which results in lower use of machinery and equipment, and in addition, these do not determine the final outcome of the process.

Owing to the geological and mining conditions (including the construction of deposits and natural and technical hazards), the type of rock being excavated (soft, such as salt or coal, and hard, such as metal ores) various technical solutions, including the degree and method of mechanisation of processes are used, and within these - the basic process, here being the extraction process (including the excavation process and processes / operations pertaining to the transport of the output). According to the selection criteria for technical solutions, there are three mechanisation systems in mining (two basic and additional - indirect) [11]:

- A system based on independent (autonomous) systems for the implementation of each particular unconnected operation - in excavation processes such systems are used in room & pillar excavation with conventional softrock mining using explosives, such as in the excavation of coal and some chemical raw materials, and hardrock mining such as metal ores etc. In these, mostly multiprocessor systems, each of the operations carried out in a given process is performed in series by a single specialised machine, working several pit-faces at one time. Among others, this is done in order to increase the utilisation of these machines. This requires successive movement of a given machine from process to process. The range of the process is most often limited by the effective blasting work (figure 5).

Figure 5. Multiple pit-face mechanised room & pillar system with single-function autonomous machinery in hardrock mining. (Source: www.youtube.com)

  The important feature of these mechanisation systems is the preparation of the excavation sites with the same set of autonomous machines as the subsequent excavation machinery. Machines belonging to the autonomous system reach the place of their work autonomously.
- A spatially coherent system, constituting of a single structural unit whose sub-assemblies perform individual operations. The most well-known excavation system of this type is fully-

mechanized longwall mining (mining by continuous miners or shearers) used to extract soft minerals buried in brown coal, coke coal or potassium salts. The system consists of a milling machine (planer or milling machine), a mechanised wall casing, a scraper conveyor belt with a crusher, as well as a power and control device (figure 6).

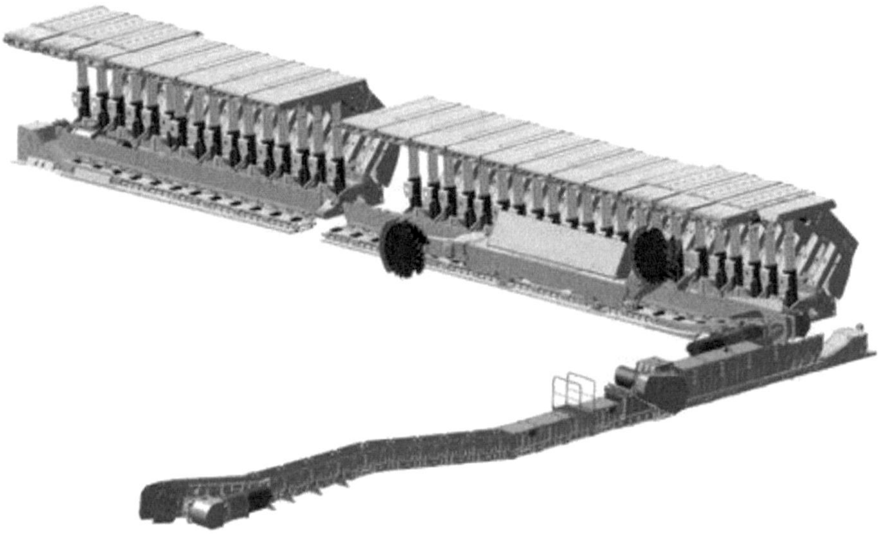

Figure 6. Continuous Miner as an example of an fully integrated mechanised system, in which sub-systems perform individual operations.

A distinguishing feature of these systems is the necessity to perform preparatory work with other technical systems (fully or partially mechanised). A separate system is also required for the supply (in elements) and assembly of the complex wall components and the reverse operation after the completion of the excavation of the section of the wall.

- Combined systems in which some of the systems are joined into a structural whole. This group includes, first and foremost, room & pillar systems using machinery such as Continuous Miners. These machines work the pit face, load the output and transport it away from the longwall to the transport vehicle (shuttle car, haul dumper or scoop loader) at several subsequent pit faces (usually 5). Such systems are used in coal mining and in the mining of rock salt, potassium and other chemical raw materials. In addition, the machinery forming part of the mechanised systems travel autonomously to the pit face and all preparatory work is performed by the same machinery.

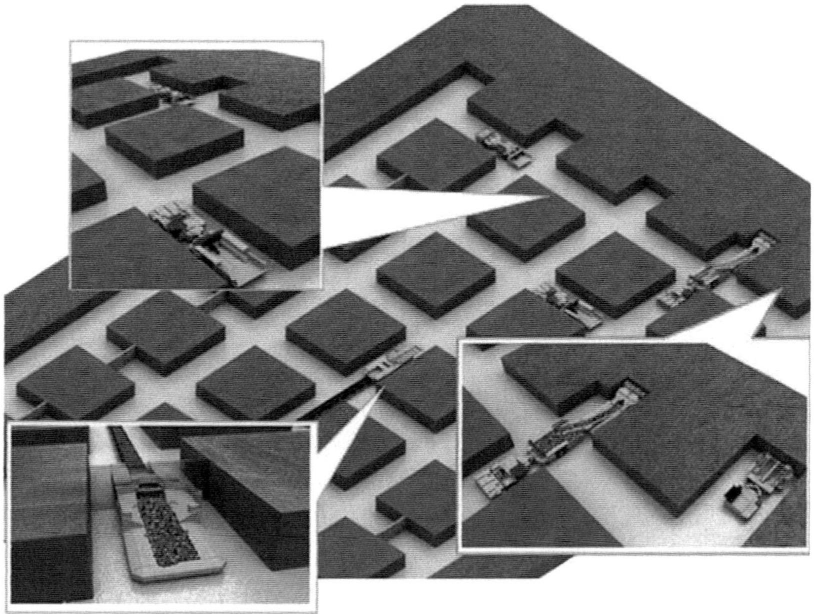

Figure 7. Multi-face system room & pillar hardcoal extraction using continuous miners and continuous haulage system (Source: www. http://monark-cat.com).

The room & pillar system (figure 7) with multi-functional machinery - a continuous miner is used without a casing (potassium salts) or with roof bolt anchors provided by as a separate function by a different machine (autonomous anchor which is wheel or caterpillar propelled).
Multifunctional coal mining machinery differs from similar machines used in the mining of potassium salts, although they perform the same functions. The range of these particular mechanization systems, the output of the mining machine - the continuous miner is usually determined by the allowable size of the exposed roof, without the need to secure it.

5. A selection of indicators and tools for the analysis of productivity applicable to mining

Indicator formulas should be set at the central level and reflect the control and implementation status of the strategic and operational objectives set. Among these are performance measures such as [12]:

- general performance,
- pit capacity,
- overall economic efficiency - defined as the ratio of sales revenue to the total number of employees,
- economic viability - defined as the ratio of sales revenue to the number of pit workers.

Productivity is defined as the relationship between received effects and the resources needed to achieve this. With productivity defined in this manner, it is possible to employ different groups of indicators in the analysis of both technical and economic aspects [13]. Some indicators of productivity analysis used in mining include [12]:

- productivity index of fixed assets (for example, daily output from a single wall in which mineral extraction is performed by a single set of autonomous machines,
- utilisation rate of machinery and equipment,
- technical and technological parameters of machines and equipment,

- an indicator linking the operational time and technical and technological parameters of machinery and equipment,
- technical equipment indicator,
- energy equipment indicator,
- indicator of technical and operational readiness of machines and equipment in excavations,
- index of utilisation of the technical capacity of machines and equipment in excavation works,
- capacity utilisation index of a wall.

Owing to the different mineral extraction technologies and different mining engineering systems, the indicators mentioned above have different forms adapted to specific processes and technical devices.

For example, the degree to which the excavator bucket or scoop capacity is utilised or the utilisation of the capacity of dump trucks used in open pit mines [8].

With respect to underground machinery (eg shaft hoists or main belt conveyors), which work "in one setting" under stable conditions, have performance indicators adapted to their conditions [14]. It must be noted that the definition the author uses for the term OEE indicator with relation to shaft lifts, differs from that used in the manufacturing industry.

A number of relevant measurements have been used in the past in Polish underground coal mining, where the mechanised, longwall mining system dominates. These have been described in the past [15], and include:

- the continuity of mining / mining rate (the ratio of mining time to longwall shearer/cutter's work on the wall),
- longwall shearer speed utilisation ratio (ratio of average cutting speed to maximum, nominal cutting speed of longwall shearer/cutter),
- indicator of the uniformity of the operation of the longwall shearer (the ratio of the minimum time of one extraction to the maximum time of one extraction),
- the utilisation of the longwall shearer in a 24-hour period (the ratio of minutes worked in terms of mining / loading of waste, to a 24-hour period expressed in minutes -1440),
- the number of minutes of failure per 1000Mg of mining (so-called breakdown rate) in daily, monthly, quarterly and annual periods,
- performance index of production of mining shifts (the ratio of the time of availability of shifts minus the duration of unplanned breaks not provided for in terms of availability, expressed in %).

In the past, calculation of these indicators was abandoned because of the manual, labor-intensive methods necessary to measuring some of them. Nowadays it is possible to calculate them automatically, i.e. in the systems of monitoring processes and equipment offered by FAMUR S.A. (E-mine), Elgór-Hansen (eg Mine-view) or other manufacturers.

Today, for the needs of its clients FAMUR S.A. offers and delivers to its customers - users reports on the effectiveness of the mining machines with the time structure shown below, within the E-mine system.

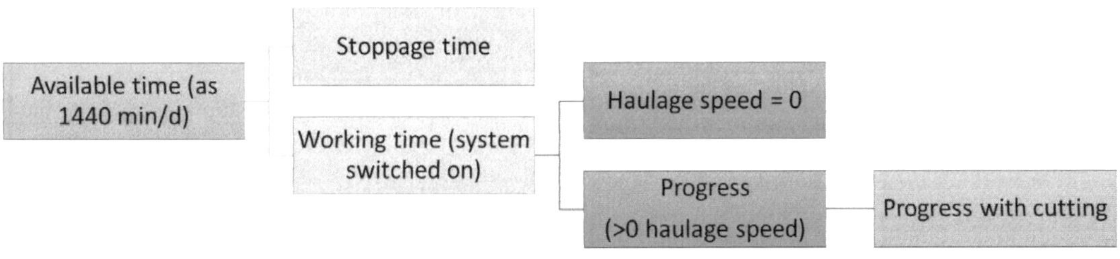

Figure 8. Structure of utilisation of continuous miner as reported within the FAMUR S.A. E-mine system.

The presented model illustrates information available in the E-mine system with regards to the loads of the electrical motors on the longwall shearer and of the speed with which it moved, as well as the various time definitions that were adopted. Also, at the request of one of the clients, the utilisation of the working time was also shown (according to [6, 15]).

Other possibilities are still being sought to identify and apply more effective measures in assessing the efficiency of mining machinery management. At present, among other things, the ratio of workers (pit or total) to one unit of production, eg using the long wall, as the measure of the efficiency of the whole mine process system (not just the basic process).

6. OEE indicator in assessing the efficiency of machinery used in the coal mining

Manufacturing processes in a properly designed production system are organised within the necessary space, which does not limit or inhibit the process. The processes for enrichment or purification processes, which may only sometimes form part of the operational processes in coal production, are in some ways similar to manufacturing processes. Therefore, it would theoretically seem possible to use the OEE to evaluate the use of processing equipment. However, the technological characteristics of mineral enrichment processes and the processing properties of technical systems make the index of equipment utilisation not fully consistent with the definition of OEE.

Mine extraction processes are much more influenced by natural factors and the resulting constraints. There is also a spatial limitation resulting from the characteristics of the deposit, the selection area or the technical and economic range of the machinery and equipment (eg the margin of profitability of transport by means of tyred vehicles or the length of the power cord). This introduces a different working time structure similar to intermittent transport. Changes resulting from speed restrictions, direction changes, or stopping machine ('red light').

While in machine production processes quality assessment is performed in quality control processes (including in-process and final quality) as well as in terms of customer complaints based on relatively easy to define indicators, defining quality in the mining process (including extraction and transport operations), is more difficult. Taking into account typical quality parameters in mining (eg ore crushing or the quality of raw coal), only to a certain extent do they depend solely on the working conditions of the available equipment. In the coal-bed wall system, the layer of false roof caving into the excavation can change its thickness by influencing the variable assessing the quality of the output. The increase in thickness of the deck beyond the technical capabilities of the applied technical system causes the loss of minerals, which in the mining industry is also part of the quality of the process. Also, the adoption of a quality measurement standard (eg ore crushing or calorific value / carbon calorific value) is not consistent with the OEE indicator.

Defining sub-indices (availability, use, quality) other than the OEE method described will result in a different quality indicator that is not comparable to OEE known from the manufacturing process. Emphasising that the OEE is only one of the KPIs and as such may be taken as a starting point for developing a new performance index covering the specific nature of the processes of hard coal mining.

7. Summary - the OEE indicator and hard coal mining processes

The basic criterion for the effectiveness of the production process is the economic result. Therefore, the search for ways to improve this result must cover all areas, and in fact, all the processes performed within the enterprise. To this end, diagnostic tools that are appropriate for the system of processes being investigated and the conditions of the system's implementation are necessary. These indicators should also allow for the comparison of the efficiency of similar processes in different manners (using different technical systems) under comparable conditions. Therefore, the OEE indicator for a relatively homogeneous group of production processes is a functional tool.

When analysing the effectiveness of using OEE indicators in hard coal mining, the following recommendations may be formulated:

- The application of diagnostic tools, methods and measurement of performance parameters, is necessary.
- The adaptation of existing industrial process diagnostics tools of one kind to another may prove pointless.
- The adaptation of existing tools and methods of diagnosing processes of one kind to another industrial type cannot consist of tailoring the process to these tools.
- The selection and application of methods and tools of process diagnosis requires knowledge of the structure of these processes and the sociotechnical systems implementing these processes.
- The OEE indicator in its definition and design is not suitable for assessing the efficiency of equipment used in the extraction process.
- Defining sub-indices (availability, use, quality) other than the OEE Indicator described will result in a different quality indicator that is not comparable to OEE as known from production processes.
- It seems appropriate to revisit the KPI indicators known and applied in the Polish mining industry in the past.

The comments presented are reflections and aim to contribute to a discussion, which is without a doubt necessary, in order to develop effective and relatively simple indicators for the assessment of coal mining process efficiency.

References:

[1] Goldratt E M 2007 Cel I: Doskonałość w produkcji (Warszawa: MintBooks) pp 383
[2] Bogacz P and Migza M 2016 Zastosowanie Lean Six Sigma w doskonaleniu procesów produkcyjnych w przemyśle wydobywczym *Inżynieria Górnicza* **4** pp 16-20
[3] Korski J 2011 Lean Management w przedsiębiorstwie górniczym *Wiadomości Górnicze* **1**
[4] Jonek-Kowalska I and Tchórzewski I 2016 Wskaźniki efektywności wykorzystania maszyn i urządzeń w górnictwie węgla kamiennego – krytyczne podejście do unifikacji i normalizacji *Inżynieria Górnicza* **4** pp 36-40
[5] Mazurek W 2014 Wskaźnik OEE – Teoria i praktyka *website: www.neuron.com.pl* (accessed:27.05.2017)
[6] Korski J 2017 Czy podejście procesowe może podnieść efektywność polskiego górnictwa węgla kamiennego? *Proc. on XXVI Szkoła Eksploatacji Podziemnej 22-24.02* (Kraków)
[7] Dauber C and Bendrat M 2014 Key performance indicator – a tool to assess ICT applications in underground coal mines *The Southern African Institute of Mining and Metallurgy*, (SOMP Annual Meeting) pp 1-20
[8] BHP *Billiton Annual Report 2014*
[9] Durlik I 1996 Inżynieria zarządzania : strategia i projektowanie systemów produkcyjnych. Cz. 1 - Strategie organizacji i zarządzania produkcją (Placet, Warszawa)
[10] Pasternak K 2005 Zarys zarządzania produkcją (Warszawa: Polskie Wydawnictwo Ekonomiczne)
[11] Klich A [ed.] 1999 Maszyny i urządzenia dla inżynierii budownictwa podziemnego. Wyrobiska korytarzowe i szybowe w górnictwie (Katowice: „Śląsk" Sp. z o.o. Wydawnictwo Naukowe)
[12] Dubiński J and Turek M 2014 Wzrost produktywności i bezpieczeństwa pracy w kopalniach szansą na funkcjonowanie i rozwój górnictwa węgla kamiennego w Polsce *Przegląd Górniczy* **4** (Katowice) pp 1-8
[13] Tobór-Osadnik K and Wyganowska M and Korski J 2016 Reasons of problems of the polish hard coal mining in connection with restructuring changes in the period 1988-2014 *Resource Policy* **48** pp 25-31 (doi: 10.1016/jresourpol.2016.02.005)

[14] Chadha R 2014 Mine improves operation through OEE and waterfall analysis *website: www.quality progress. com* (accessed: 20.05.2017)
[15] Matuszewski J and Masarczyk J 1976 Metody usprawniania organizacji produkcji w ścianach węglowych (Katowice: Wydawnictwo „Śląsk”)

IOP Conf. Series: Materials Science and Engineering **268** (2017) 012011 doi:10.1088/1757-899X/268/1/012011

Disposal of coal combustion wastes in the hydraulic backfill process

Piotr Pierzyna

Faculty of Mining and Geology, Silesian University of Technology, Akademicka 2, 44-100 Gliwice, Poland

piotr.pierzyna@polsl.pl

Abstract. This article presents the results of studies regarding the physical properties of selected combustion by-products (CCPs) currently produced in the energy production industry. These properties have been compared with the requirements of the technologies applied in the Polish underground mines. The article gives special consideration to the application of the products in the hydraulic backfill technology. The possibility of using bottom-ashes and slags was considered. The amount of CCPs disposed in Polish hard coal mines is approximately 1.1 million Mg and the tendency is decreasing. In the past two years, approximately 100-150 thousand Mg of CCPs was used in the hydraulic backfill technology. The percentage of the fraction smaller than 0.1 mm is determining for the possibility of using a given type of CCPs in the backfill material. This practically excludes the possibility of using any fly ashes in that technology. In slags from conventional boilers and bottom ashes from fluidized bed boilers the fraction below 0.1 mm constitutes 25% of the total at maximum, which allows for their use in the materials used in hydraulic backfill as a component comprising from 30% to 60%, respectively. Slags (10 01 01) are characterized by the lack of bonding properties, which, in case of open backfill systems that are exposed to atmospheric conditions, constitutes an advantage in comparison to bottom ashes (10 01 24), which in turn definitely exhibit bonding properties. The solution of the problem of using bottom ashes is their supply and application on a current basis.

1. Introduction

Energy production wastes are often called combustion by-products. In global nomenclature, they are most often referred to as Coal Combustion Products (CCPs), or – less often – as Coal Combustion Wastes (CCW) or Coal Combustion Residuals (CCR). The wastes include all kinds of slags from conventional boilers and ashes from fluidized bed boilers, fly ashes and slag and ash mixtures as well as flue gas desulfurization products.

The supreme legal act regulating waste management is the Act on Wastes dated December 14th, 2012 (2013 Journal of Laws pos. 21, with amendments), while the division of wastes and the indication of hazardous wastes has been given in the Ordinance of the Minister of Environment dated December 9th, 2014, regarding the catalogue of wastes (2014 Journal of Laws pos. 1923). Energy production wastes have been included in group 10: Wastes from thermal processes, subgroup 01: Wastes from power plants and other plants producing energy by combustion of fuels (excluding group 19). The Act referred to above allows for 13 recovery processes provided in its appendix 1 (R1-R13).

Backfill systems in Polish coal mines are systems as understood by the Ordinance of the Minister of Environment dated August 27th, 2014 (2014 Journal of Laws, pos. 1169) regarding the types of systems which may cause significant pollution of elements of environment or the environment as a whole.

The recovery of wastes and thus their disposal in the mining plants is made possible by the Ordinance of the Minister of Environment dated May 11th, 2015 regarding the recovery of wastes outside systems and devices. The ordinance specifies the types of wastes and the conditions for their recovery outside of systems or devices in the R3, R5, R11 and R12 processes, while in mining, the R5: Recycling or recovery of other inorganic materials process is applied.

The beginnings of the deposition of energy production wastes in Polish underground mines reach the 1950s. At that time, the 'Borynia' coal mine did the first large-scale attempt – although unsuccessful – to apply the wastes as a hydraulic backfill material. At that time, trials using fly ashes in fire prevention were conducted in Poland. Simultaneously, similar trials were conducted in former Czechoslovakia, in the Ostrava-Karviná basin (current area of Czech Republic). In later years, using the experience of British engineers, power production wastes were used to fill shallow voids and were applied pneumatically [1]. The introduction of slurry technology and the application of emulsifiers, especially ash and water, constituted a breakthrough which allowed for a mass-scale application in Polish underground mines. The first trial using the slurry technology was conducted in the 'Generał Zawadzki' underground hard coal mine. [2] Studies and technologies using power production wastes in the Polish underground mines flourished in the nineties, when Poland started functioning as a market economy and the mining industry was restructured by, among other things, seeking costs.

The contemporary technological advances in the power production industry resulted in the development of fluidized bed boilers and thus a new quality of wastes. An interesting direction in the management of power production wastes, which was first developed in the nineties and is still developed in the Polish underground mining, was the production of different types of bonding agents based on this type of wastes. The PPH Utex Sp. z o.o. company from Rybnik was the precursor in the usage of power production wastes.

The use of power production wastes in underground mining technologies allows for [3]:

- effective recovery and mass-scale management of wastes,
- limiting the deposition on the surface and the related costs – ecological aspect,
- limiting the fire hazard and the related methane and coal dust explosion hazards – improvement in terms of occupational safety,
- simplification and improvement of ventilation conditions – financial savings,
- different types of applications that were not profitable beforehand – by using the wastes in various bonding agents.

Power production wastes, both directly and as a component of bonding materials, may be used in underground mining in technologies such as [4]:

- additional gob grouting and extinguishing fires,
- solidifying and hydraulic backfill,
- filling of unnecessary headings and chambers,
- backfill plugs,
- liquidation of shallow voids caused by former exploitation,
- liquidation of shafts,
- liquidation of boreholes,
- reconsolidation of fallen roof and building of artificial roof.

The use of power production wastes also allows for the disposal of other materials which are present in coal mines on mass-scale, such as salt water, barren rock and flotation wastes [5, 6, 7].

Additionally, power production wastes may constitute a component of mineral bonding agents used in mines for:

- The production of shotcrete,
- Building explosion-proof plugs,
- Building roadside packs.

Table 1 presents the amounts of combustion by-products disposed in the years 2014-2016 in division into: KW SA (currently PGG Sp. z o.o.), KHW SA (currently PGG Sp. z o.o.), JSW SA and a group of remaining mines [1].

Table 1. Amounts of combustion by-products disposed in the Polish coal mines in the years 2014-2016 [1].

Coal mining companies and remaining mines		2014	2015	2016
			(Mg)	
Disposal of combustion by-products in mines – total, including:		**1 180 451**	**1 179 784**	**1 142 733**
KW SA		643 967	470 025	352 048
KHW SA		164 523	146 239	156 079
JSW SA		340 639	302 745	360 606
Remaining mines		31 322	260 774	274 000
method of use	hydraulic backfill	**23 781**	**150 380**	**99 785**
	remaining uses	**1 156 670**	**1 029 404**	**1 042 947**

All industrial wastes, especially resulting from power production, the so-called combustion by-products, that are to be used in underground mining technologies, must fulfil specified requirements relating to their physical and chemical properties as well as their radioactivity. The fulfilment of these requirements allows for their safe use. As of today, the mining industry applies two standards as far as the coal combustion products are concerned:

- PN-G 11011:1998 entitled 'Materials for solidified backfill and gob grouting. Requirements and testing',
- PN-G 11010:1993 entitled 'Hydraulic backfill materials. Requirements and testing'.

As it may be noted, the above standards only apply to three of the technologies referred to above, namely: gob grouting, solidified backfill and hydraulic backfill. In case of the remaining technologies referred to above, the designers often refer to the above standards in full or to a part thereof.

The results presented in the article were obtained in tests conducted in line with standards corresponding to given parameters: PN-G 11010:1993, PN-G 11011:1998 and PN-EN 15309:2010.

2. Standard requirements of the mining industry relating to physical properties

The requirements of the PN-G 11011:1998 standard regarding the physical values relate to the materials themselves and their mixtures with water (after mixing) that is, the liquid and solidified materials.

The requirements regarding the materials 'themselves' are common for the backfilling and grouting technologies and concern two parameters, namely:

1. the density, which is to be higher than 1.0 Mg/m^3, for the materials not to be subject to sedimentation;
2. the highest allowable grain size, which may not exceed 1/3 of the transport pipeline diameter, as described in the former part of this article.

Combustion by-products generally fulfil the criteria required by the standard. Only the use of fly ashes which contain large amounts of so-called microspheres, the density of which is approximately 0.5 Mg/m^3, may pose a problem. Usually, however, the percentage of microspheres in fly ashes is low and reaches up to about 2% [8].

Requirements regarding the physical properties of materials used in hydraulic backfilling as specified by the PN-G 11010:1993 standard have been given in table 2.

Table 2. Requirements regarding the physical properties of materials used in hydraulic backfilling as specified by the PN-G 11010:1993 standard.

Material class	Content of grains above 0.1 mm, max. %	Max grain size mm	Max. compressibility at 15 MPa %	Min. water permeability cm/s	Leaching capability max. %	Min. bulk density kg/dm³	Content of evident plant parts
I	10		5	0.007			
II	15	60.0	10	0.002	20 [1)]	1.3	should not contain
III	20		15	0.0004			

[1)] Materials which exhibit leaching higher than 20% within 30 minutes and smaller than 20% within 15 minutes may only be used in backfill systems up to 2.5 km in length.

As it may be noted, the backfill standard distinguishes three classes of backfill material, determined based on the grain size above 0.1 mm, the water permeability and – most of all – the compressibility, which may not exceed 15%.

3. Grain size composition – a factor determining the usability of combustion by-products in given technologies

Grain size composition is the basic physical quality of power production wastes and is decisive for the possibility of their application in mining technologies. Grain size is the division into fly-ash and slags from conventional boilers (in line with BN - 79/7622-09) as well as bottom-ash from fluidized bed boilers. The grain size composition has a significant impact on parameters such as: compressibility, water permeability, sedimentation speed or propagation [9]. The grain size composition of fly-ash, slags, bottom ashes, filling sand and barren rock has been presented in figure 1.

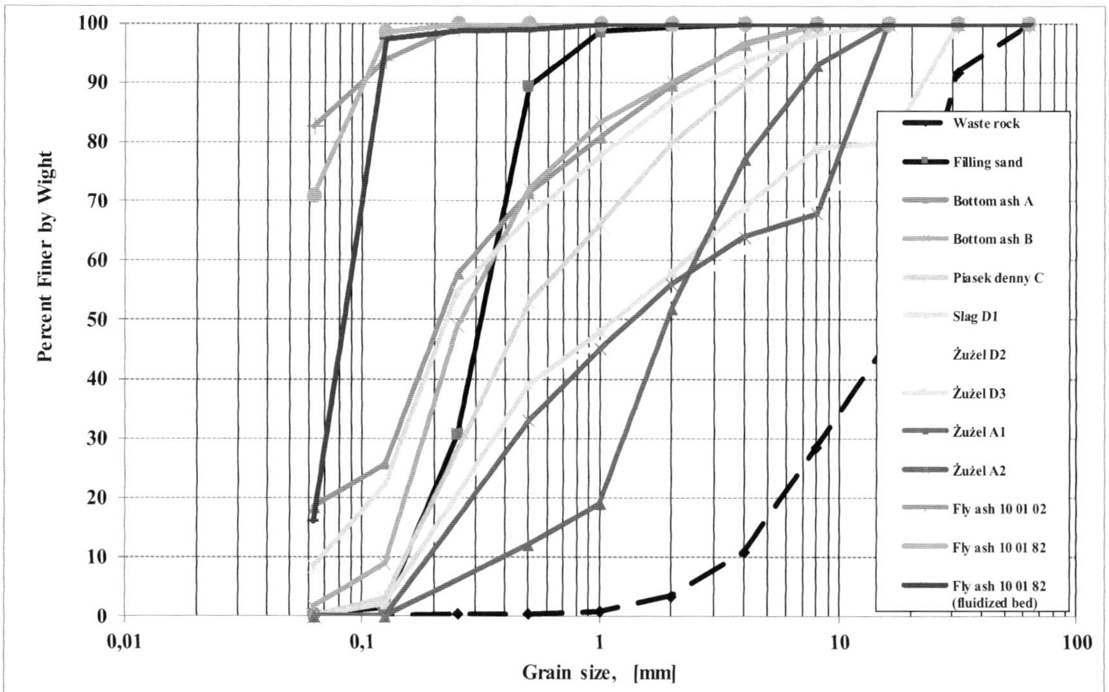

Figure 1. The grain size composition of fly-ash, slags, bottom ashes, filling sand and barren rock.

Fly ashes (10 01 02 i 10 01 82) are wastes characterized mostly by the grain sizes below 0.1 mm, which comprise over 70% of the total. Usually this value reaches approximately 90%. As a result, the use of these ashes poses a problem in view of their use in the backfill. In line with the PN-G 11010:1993 standard, the maximum percentage of grains below 0.1 mm may amount to 20%. Thus, the theoretical percentage of ashes in the backfill material will be low. Besides, giving consideration to the experience gained in tests conducted by the Institute of Mining of the Silesian University of Technology related to fly-ash, it should be assumed that the sedimentation of ash particles is slow, which causes their transportation outside the backfilled area and the silting of the mine drainage system. On the other hand, the slow sedimentation of fly-ash is desirable in case of the gob grouting technology or in case of the technology of filling various kinds of underground headings or shallow voids.

While analysing the graining curves of other combustion by-products – slags and bottom ashes – it may be concluded that the curves fall in between the curve of typical filling sand and the graining curve of crushed barren rock. It should be mentioned that sand, often mixed with crushed barren rocks such as slates, is a traditionally used backfill material. While analysing the grain size composition of slags (10 01 01), it may be concluded that the percentage of the fraction below 0.1 mm is lower than 20%. Numerous studies conducted by the author indicate that the percentage is mostly below 5%. Bottom ashes, on the other hand (10 01 24) contain 25% of grains below 0.1 mm in size, and usually the percentage is below 10%. The above analysis exhibits the potential to apply slags and bottom ashes in materials used in hydraulic backfilling technology. Similar results of the CCPs grain size composition have been exhibited in the work [10]. Further part of the article presents test results of slags and bottom ashes in view of the requirements of the PN-G 11011:1998 'Materials for hydraulic backfilling…' standard.

4. Compressibility of backfill mixtures containing slag or bottom ashes

The backfill standard, as mentioned before, distinguishes three classes of backfill material, determined mostly by its compressibility, which may reach the value of 15% at a pressure of 15 MPa at maximum. The compressibility test results of a mixture of filling sand with bottom ash or slag at different mass fractions and at a pressure of 15 MPa have been presented in figure 2.

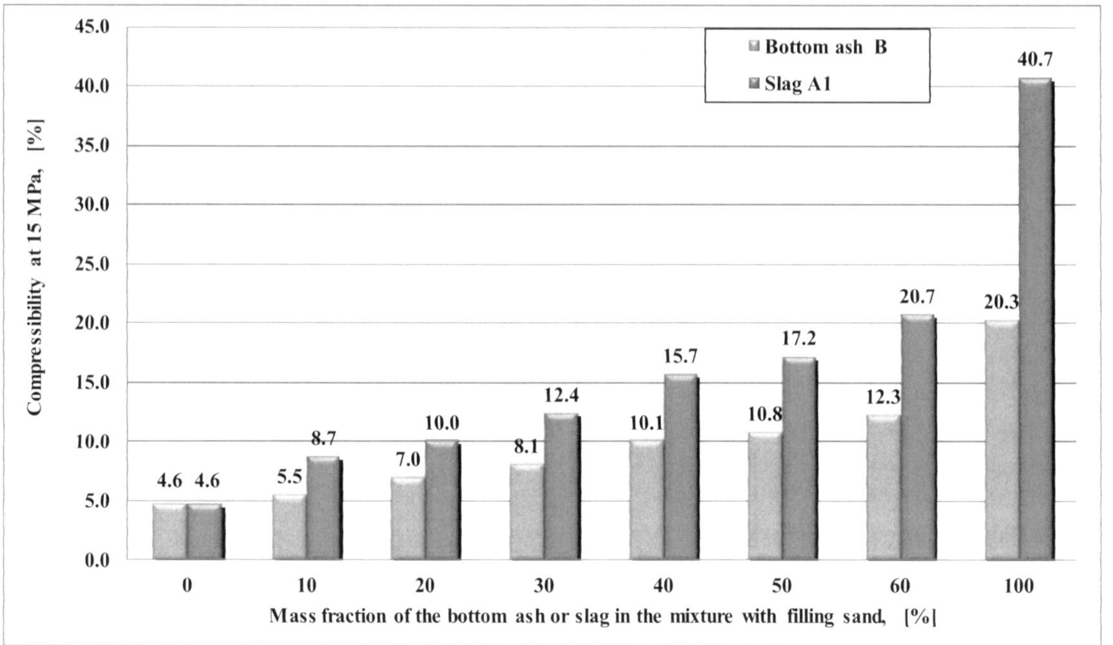

Figure 2. The compressibility test results of a mixture of filling sand with bottom ash or slag at different mass fractions and at a pressure of 15 MPa.

While analysing the presented test results, it should be noted that bottom ash used as an example exhibits compressibility twice as low as that of slag. In case of the bottom ash used as the example it is approximately 20%, while in case of slag it reaches 40%. The compressibility results both for the bottom ash and for the slag were obtained in line with the PN-G 11010:1993 standard as a 'standalone' backfill material. Both of the above are non-classed materials.

Usually mines require class I backfill materials exhibiting compressibility up to 5%. In that case, it is not possible to use slag or bottom ash in a mixture with filling sand. Their percentage would have to be kept below 10%. Polish mines do not dispose of systems in which such a precise dosing of an additional component to the filling sand would be possible. Besides, the profitability of the solution would be negligible.

In case of class III backfill material (compressibility up to 15%), however, the maximal fraction of bottom ash in the mixture with filling sand may reach up to approximately 50-60%, while the percentage of slag may be half the value, that is at a level of approximately 30%.

The test results and the analysis presented above indicate a practical possibility to use bottom ash or slag in a mixture with filling sand to obtain a class II or class III backfill material in line with the requirements of the PN-G 11010:1993 standard.

5. Water permeability of backfill mixtures containing slag or bottom ashes

Water permeability is yet another parameter which distinguishes the classes of the backfill material. The minimal water permeability required by the standard is 0.0004 cm/s. Water permeability in case of materials hydraulically placed in headings, as in the case of the backfill, constitutes the possibility of water bleeding from the material. The higher the water permeability of materials used in underground workings, the lower the pressure on the isolating dams, the quicker the yielding capacity achieved and the lower the water flooding hazard for the longwall's working area and the miners.

Water permeability test results of a mixture of filling sand with bottom ash or slag at different mass fractions have been presented in figure 3. [own source].

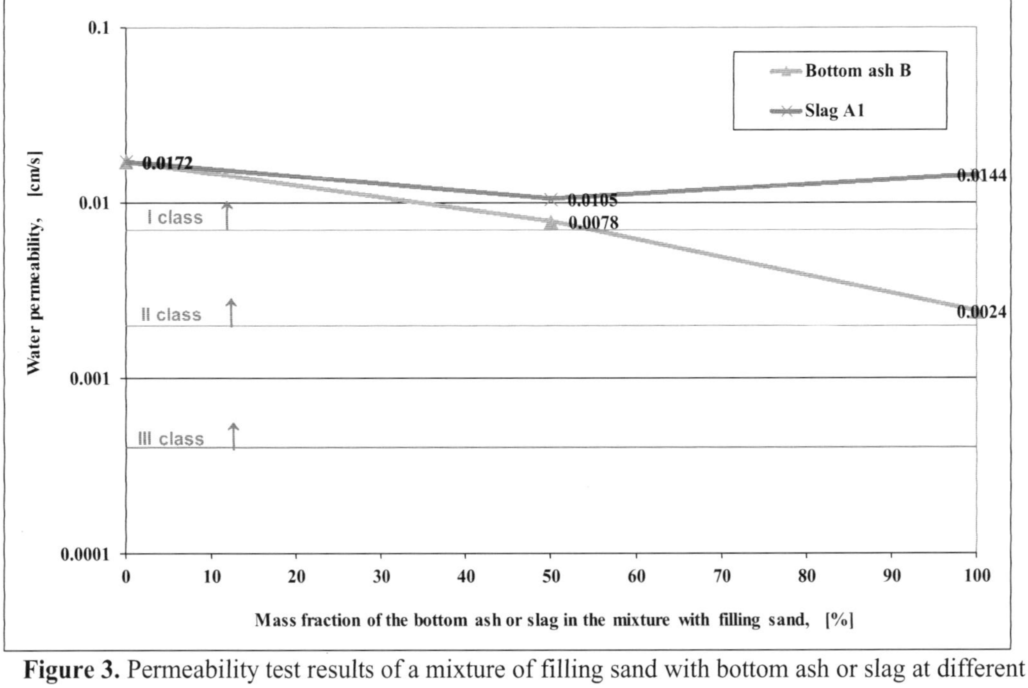

Figure 3. Permeability test results of a mixture of filling sand with bottom ash or slag at different mass fractions.

While analysing the presented test results, it should be noted that bottom ash used as an example exhibits water permeability lower by one order of magnitude, as compared to the slag. In case of the bottom ash used as the example it is approximately 0.0024 cm/s, while in case of slag it reaches 0.0144 cm/s. The obtained water permeability results, both for the bottom ash and for the slag, classify the materials as class II or class I backfill materials in line with the PN-G 11010:1993 standard. Considering the above results of water permeability tests in view of the PN-G 11010:1998, it should be concluded that bottom ashes and slags may be used even as standalone backfill materials. As it has been mentioned earlier, however, due to their compressibility, their fraction is limited to approximately 60% in case of bottom ash and to 30% in case of slag.

6. Bonding properties of slags and bottom ashes

While considering the possibility of applying slags and bottom ashes in hydraulic backfilling, one should note their bonding properties. In slags and in bottom ashes the properties depend on the correlations between CaO-Al_2O_3-SiO_2-Fe_2O_3 [11]. The bonding properties of a given mineral material may be determined eg. by calculating the so-called hydraulic modulus (M_h). The hydraulic modulus is a ratio of amount [% of weight] of the calcium oxide to the sum of the oxides of silicon, aluminium and iron [12, 13]. The hydraulic modulus determines the type of the bonding agent, mainly dividing them into hydraulic and air types. For example, in case of Portland cement, the hydraulic modulus is between 1.7 and 2.4 [14]. Table 3 presents the hydraulic modulus of examples of bottom ashes and slags. The individual components were determined based on the wavelength dispersive X-ray fluorescence spectrometry method in line with the PN-EN 15309:2010 standard.

Table 3. The value of the hydraulic modulus for the bottom ash and slags serving as the example.

Waste (CCPs)	CaO (%)	Al_2O_3 (%)	SiO_2 (%)	Fe_2O_3 (%)	M_h (-)
Bottom ash A	20.42	15.93	35.03	2.89	0.38
Bottom ash B	15.99	16.19	41.58	2.97	0.26
Bottom ash C	15.06	18.03	44.13	3.32	0.23
Bottom ash D	10.22	17.1	54.7	3.96	0.13
Slag A1	0.99	20.31	43.53	6.06	0.01
Slag A2	2.90	19.53	54.17	10.92	0.03
Slag D1	4.71	20.71	47.55	6.34	0.06
Slag D2	2.43	16.12	50.74	14.73	0.03

The values of the hydraulic modulus obtained in calculations indicate that slags (10 01 01) do not exhibit bonding properties. It is different in case of bottom ash (10 01 24), which exhibits such properties. This is disadvantageous, as the ash may be solidified in the backfill material tanks located at the surface, as they are open and affected by atmospheric conditions, especially precipitation and moisture. In case of using bottom ash exhibiting bonding properties in hydraulic backfill technology, it is important for it to be used on a current basis, immediately after the delivery. The bonding properties of bottom ash, their compressibility after 7, 14 and 28 days of seasoning in humid air conditions [own work] have been presented in figure 4.

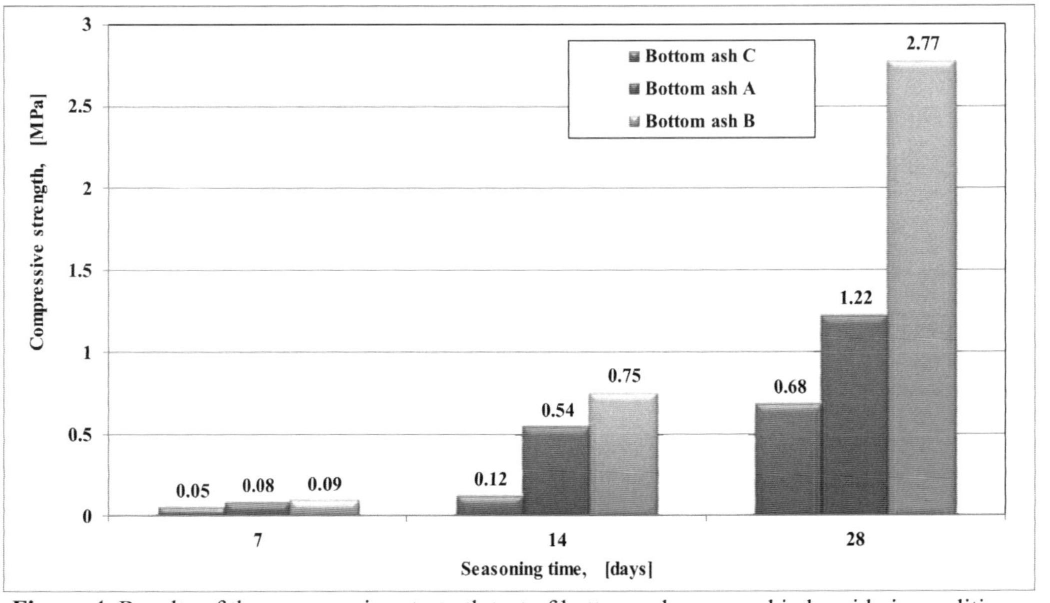

Figure 4. Results of the compressive strength test of bottom ash seasoned in humid air conditions, exhibiting their bonding properties [own source].

7. Summary

The amount of CCPs disposed in hard coal mines is approximately 1.1 million Mg and the tendency is decreasing. In the past two years, approximately 100-150 thousand Mg of CCPs were used in the hydraulic backfill technology.

The percentage of the fraction below 0.1 mm is decisive for the possibility of using a given type of combustion by-products in the backfill material. This practically excludes the possibility of using any fly ashes in that technology. On the other hand, in slags from conventional boilers and bottom ashes from fluidized bed boilers the fraction below 0.1 mm constitutes 25% of the total at maximum (usually below 10%), which allows for their use in the materials used in hydraulic backfill.

While considering the compressibility and possibility of using slag or bottom ash in mixture with filling sand it may be noted that in case of:

- **Class I** – the possibility is low. The percentage would have to be kept below 10%, which may be considered unprofitable. Moreover, mines do not own systems allowing for such a precise proportioning of a component added to the filling sand.
- **Class III** – the maximal percentage of bottom ash in the mixture with filling sand may reach up to approximately 60%, while the percentage of slag may be half the value, that is at a level of approximately 30%.

Water permeability of slags or bottom ash and their mixtures with backfill sand is above the lower threshold given by the standard, that is 0.0004 cm/s. Slags allow for achieving higher water permeability than it is in the case of bottom ash. This is also valid in case of their mixtures with filling sand.

Slags from conventional boilers (10 01 01) are characterized by a lack of bonding properties, which, in case of open filling systems that are exposed to atmospheric conditions, constitutes an advantage in comparison to bottom ashes from fluidized bed boilers (10 01 24), which in turn definitely exhibit bonding properties.

The solution of the problem of using bottom ashes is their supply and application on a current basis. One should note, however, that in case of longer downtime related to no advancement of the backfilled longwall, the mine should be equipped with an additional location that allows for the application of the backfill material from a tank.

References

[1] Drobek L, Kanafek J and Pierzyna P 2016 Zagospodarowanie UPS w kopalniach węgla kamiennego stan aktualny, prognoza zużycia na lata 2016 - 2020, aspekty technologiczne i środowiskowe *XXIII Międzynarodowa konferencja Popioły z energetyki* ed. A Kornacki. (Warszawa: Polska Unia UPS) pp 225-37

[2] Piotrowski Z and Mazurkiewicz M 2006 Chłonność doszczelnianych zrobów zawałowych *Górnictwo i Geoinżynieria* **30/3** pp 37–46

[3] Chugh Y P, Biswas D and Deb D 2002 Undergroung placement of coal processing waste and coal combustion by-products based paste backfill for enhanced mining economics (Southern Illinois University (US))

[4] Palarski J 2012 *Odzysk i unieszkodliwianie odpadów w górnictwie podziemnym* (Gliwice: Silesian University of Technology)

[5] Mahlaba J S, Kearsley E P, Kruger R A and Pretorius P C 2011 Evaluation of workability and strength development of fly ash pastes prepared with industrial brines rich in SO4= and Cl− to expand brine utilisation *Miner. Eng.* **24** pp 1077–81

[6] Strozik G 2017 Reduction of Saline Waters Discharge from Coal Mines Through Filling and Sealing of Underground Voids *World Sci. News* **72** pp 10–25

[7] Plewa F and Pierzyna P 2008 Wykorzystanie ubocznych produktów spalania do utylizacji wód kopalnianych *Polityka Energetyczna* **11/1** pp 361–73

[8] Yu J, Li X, Fleming D, Meng Z, Wang D and Tahmasebi A 2012 Analysis on Characteristics of Fly Ash from Coal Fired Power Stations *Energy Procedia* **17** pp 3–9

[9] Plewa F and Mysłek Z 2010 Analiza wpływu doszczelniania zrobów zawałowych odpadami drobnoziarnistymi na zmniejszenie deformacji powierzchni w kopalniach węgla kamiennego *Przegląd Górniczy* **66/12** pp 92–5

[10] Senapati P K and Mishra B K 2015 Bulk Hydraulic Disposal of Highly Concentrated Fly Ash and Bottom Ash–Water Slurries *Part. Sci. Technol.* **33** pp 124–31

[11] Argiz C, Sanjuán M Á and Menéndez E 2017 Coal Bottom Ash for Portland Cement Production *Adv. Mater. Sci. Eng.* pp 1–7

[12] Wirska-Parachoniak M 1968 Z historii wiążących materiałów budowlanych *Ochr. Zabyt.* pp 17–23

[13] Sitko J 2016 Modernizacja technologii zagospodarowania odpadów hutniczych *Systemy Wspomagania w Inżynierii Produkcji*

[14] Nakano T, Yokoyama S, Uchida S and Maki I 2007 Badania podstawowe związane z produkcją cementu z popiołu ze spalania miejskich odpadów stałych. Cz.1: Charakterystyka popiołów i cementów specjalnych *Cement Wapno Beton* pp 187–92

The impact of changes in the rheological parameters of fine-grained hydromixtures on the efficiency of a selected industrial gravitational hydraulic transport system

Marcin Popczyk

Faculty of Mining and Geology, Silesian University of Technology, Akademicka 2, 44-100 Gliwice, Poland

marcin.popczyk@polsl.pl

Abstract: Polish hard coal mines commonly use hydromixtures in their fire prevention practices. The mixtures are usually prepared based on mass-produced power production wastes, namely the ashes resulting from power production [1]. Such hydromixtures are introduced to the caving area which is formed due to the advancement of a longwall. The first part of the article presents theoretical fundamentals of determining the parameters of gravitational hydraulic transport of water and ash hydromixtures used in the mining pipeline systems. Each hydromixture produced based on fine-grained wastes is characterized by specified rheological parameters that have a direct impact on the future flow parameters of a given pipeline system. Additionally, the gravitational character of the hydraulic transport generates certain limitations concerning the so-called correct hydraulic profile of the system in relation to the applied hydromixture characterized by required rheological parameters that should ensure safe flow at a correct efficiency [2]. The paper includes an example of a gravitational hydraulic transport system and an assessment of the correctness of its hydraulic profile as well as the assessment of the impact of rheological parameters of fine-grained hydromixtures (water and ash) produced based on laboratory tests, depending on the specified flow parameters (efficiency) of the hydromixture in the analyzed system.

1. Introduction

Hydraulic transport of fine-grained mixtures in pipeline systems that is in use in Polish hard coal mines is most often of a gravitational type. In systems of that kind, the flow of the hydromixture is a self-controlled process, which means that the parameters of flow result from the spatial configuration of the system, especially the ratio of its length to the difference between the inlet and outlet altitudes, the diameter of the pipeline and the thickness of the mixture. Along with the increase of the length to altitude difference ratio, the flow efficiency parameters of the backfill mixture decrease, which results from the decreased flow rate [3]. To ensure stable flow parameters of the mixture, the system should be characterized by a correct hydraulic profile, that is, a profile in which the pressure curve does not cross the hydraulic profile of the system. In cases where the system is characterized by a disadvantageous hydraulic profile, the flow of the mixture in the system is disrupted and the achievement of the assumed flow parameters becomes impossible. In such a case the system contains negative pressure areas, which may lead to cavitation resulting in the decreased life of the pipeline. In case of using water and ash mixtures, the rheological properties such as viscosity and flow limit of the mixture have a significant impact on the parameters of hydraulic transport and should be considered in the calculations. Currently in the Polish mining industry the hydraulic gravitational transport of fine-grain mixtures is commonly used for fire prevention in the technology of additional gob grouting.

The paper presents an analysis of the geometric parameters of the system for the transport of fine-grain hydromixtures to a gob grouting location in an underground coal mine. Subsequently, an analysis of the impact of the rheological parameters of the hydromixtures on the efficiency parameters of the process was conducted.

2. Method for determination of flow parameters of fine-grain mixtures in gravitational systems

The flow of the fine-grain hydromixture in gravitational systems proceeds due to the force of gravity. It is assumed that the movement of the hydromixture is described by a modified D. Bernoulli equation, while the unitary energy losses are determined empirically. The notion of the modified Bernoulli equation should be understood as a generalized Bernoulli equation used for the description of the actual movement of liquid in pipeline, containing additional corrective factors, in which the density of the liquid has been replaced with the kinetic density of the hydromixture. At a specified flow rate of the hydromixture in gravitational system, the pressure at the end of the vertical string, figure 1 is approximately [4, 5]:

$$p_B = h_1 \rho_m g \eta - \Delta p_{A-B} h_1 \tag{1}$$

In the formula, the symbol ρ_m denotes the so-called kinetic density of the hydromixture, that is, the mass of solid particles and water in 1 m^3 of the hydromixture in movement. The η coefficient is a corrective factor, which was named – quite incorrectly – the hydrodynamic efficiency factor of the system.

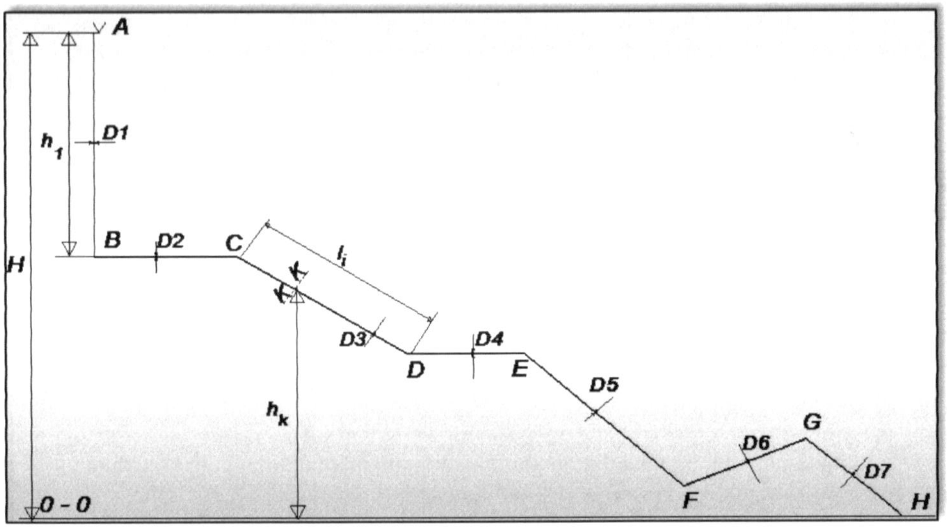

Figure. 1. Scheme of gravitational backfill installation.

Unitary energy losses in the vertical pipeline A-B were assumed to be similar to the unitary energy losses of hydromixture flow in a horizontal pipeline with a similar diameter. The generalized Bernoulli equation for the A-A and B-B sections of the gravitational system have been presented in figure 1 after assuming that the hydromixture was replaced with an adopted homogenous fluid with a density corresponding to the mean actual density of the hydromixture that is fixed at all sections and all points of the stream, is as follows:

IOP Conf. Series: Materials Science and Engineering **268** (2017) 012012 doi:10.1088/1757-899X/268/1/012012

$$h_1\rho_m g + p_A = \frac{v_m^2}{2}\rho_m + p_B + \Delta p_{A-B}h_1 \tag{2}$$

where:

$p_A = p_B$ – atmospheric pressure (barometric),

After a transformation, the following form is obtained:

$$p_B = h_1\rho_m g + p_A - \frac{v_m^2}{2}\rho_m - \Delta p_{A-B}h_1 \tag{3}$$

In practice, in gravitational systems with depth higher than 100 m, the value of the expression $p_A - \frac{v_m^2}{2}\rho_m$ is small in comparison to the $h_1\rho_m g$ value and may be neglected.

The unitary energy losses assume smaller values in a vertical pipeline than in a horizontal pipeline, subject to the assumption that the pipeline is accurately vertical. It may be thus noted that the factor may be assumed to be equal to 1. In a gravitational system, steady flow proceeds due to the fixed difference in the hydromixture levels. If the distance between the level of the outlet cross-section of the hydromixture and the level of the hydromixture in the hopper is denoted by *H*, figure 1, then, by applying the D. Bernoulli rule to the top surface and any K-K section, the following formula is obtained:

$$H + \frac{p_B}{\rho_m g} = h_K + \frac{p_K}{\rho_m g} + \frac{v_{mK}^2}{2g} + \frac{1}{\rho_m g}\sum_A^K l_{A-K}\Delta p_{A-K} \tag{4}$$

After a transformation of the formula (4), the following form is obtained:

$$\frac{p_K}{\rho_m g} = \frac{p_B}{\rho_m g} + H - h_K - \frac{v_{mK}^2}{2g} - \frac{1}{\rho_m g}\sum_A^K l_{A-K}\Delta p_{A-K} \tag{5}$$

The equation indicates that the pressure at any section of the gravitational system decreases along with the increase of the speed of the hydromixture and along with the increase in the energy losses of the flow, which assume higher values at a higher mean flow rate.

The continuous flow of the hydromixture in the pipeline is possible when the absolute pressure at each point of the pipeline does not drop below the pressure of vapour of saturated fluid in which the solid particles are transported in a given temperature

$$\left(\frac{p_v}{\rho_w g} = 0.238m\, H_2O \text{ in the temeperature of } 20^o\text{C}\right)$$

If the vaporization pressure is denoted by p_v, the condition of the flow continuity may be presented as:

$$\frac{p_{\min}}{\rho_m g} = \frac{p_B}{\rho_m g} + H - h_K - \frac{v_{mK}^2}{2g} - \frac{1}{\rho_m g}\sum_A^K l_{A-K}\Delta p_{A-K} \geq \frac{p_v}{\rho_m g} \tag{6}$$

Thus, the maximal allowable flow speed of the hydromixture at any K-K section:

$$v_{mK_{\max}} = \sqrt{2g\left(\frac{p_B}{\rho_m g} - \frac{p_v}{\rho_m g} + H - h_K - \frac{1}{\rho_m g}\sum_A^K l_{A-K}\Delta p_{A-K}\right)} \tag{7}$$

In the expression under the square root sign, the volume of the energy losses is also a function of speed. That is why while determining the maximal flow rate one needs to know the energy losses equation. In practice, the mean flow speed of the hydromixture should be within the range of:

$$v_{kr} \leq v_m \leq v_{gr}$$

The sedimentation speed (critical) is a speed below which particles are deposited at the bottom of the pipeline. To ensure the safe flow and to prevent the deposition at the bottom of the pipeline, it is assumed that the minimal value of the movement safety factor M, which is the ratio of the flow speed of the mixture in the pipeline and the critical speed, should be at least 1.3. The threshold speed is a conventional speed of hydromixture flow,above which the abrasibility of the pipeline increases significantly. The tests of pipe abrasibility in gravitational systems have indicated that the speed above which the abrasibility increases significantly is 10 ÷ 12 m/s.

The basic problem while determining the parameters of the hydraulic transport of fine-grained mixtures is constituted by the necessity of a correct determination of unitary flow drag. The knowledge of the rheological parameters is the condition of a correct determination of unitary energy losses of non-Newtonian fluids flow, which include the high-density ash and water mixtures.

The rheological parameters may be specified by determining the flow curve in the laminar flow zone and by selecting a corresponding rheological model.

The following rheological models are usually applied to describe the flow of fine-grain mixtures:

- Newtonian model

$$\tau = \tau_0 + \eta \frac{\Delta v}{\Delta y} \tag{8}$$

- Ostwald de Waele's model

$$\tau = k\left(\frac{dv}{dy}\right)^n \tag{9}$$

- Bingham's model

$$\tau = \tau_0 + \eta_p \frac{dv}{dy} \tag{10}$$

- Casson's model

$$\sqrt{\tau} = \sqrt{\tau_0} + \sqrt{\eta \frac{dv}{dy}} \tag{11}$$

where: τ – steady stress, Pa,
τ_0 – flow threshold, Pa,
η - viscosity, Pa*s,
dv/dy – shearing speed, s^{-1},
k, n – model's parameters.

Flow curves of fine-grained mixtures may be determined using the measurements of steady stresses in the function of shearing speed using rotational viscometers with coaxial measurement cylinders. Knowing the flow curve and selecting an optimal rheological model, rheological parameters and unitary drags of flow of the mixture in the laminar flow zone may be determined.

In the turbulent flow zone, where a part of the mixture is dispersed into chaotic movement of solid particles and water, it is most convenient to determine the flow parameters including the unitary drags of flow, based on the measurements of flow of mixture in laboratory systems. The measurements allow for the determination of the λ linear drags coefficient which is used in the basic dependency allowing for the determination of the flow speed in turbulent flow at a known volume of unitary flow energy losses:

$$\Delta p = \lambda \rho_m g \frac{v_m^2}{2D} \qquad [\text{Pa/m}] \tag{12}$$

The knowledge of the flow speed allows for the determination of the efficiency parameters for the flow of a fine-grain mixture:

- Volumetric flow rate of the mixture Q_m:

$$Q_m = v_m \frac{\pi D^2}{4} \qquad [\text{m}^3/\text{s}] \tag{13}$$

- Volumetric flow rate of solid particles Q_s:

$$Q_s = C_V Q_m \qquad [\mathrm{m^3/s}] \qquad (14)$$

- Volumetric flow rate of water Q_w:

$$Q_w = (1 - C_V) Q_m \qquad [\mathrm{m^3/s}] \qquad (15)$$

After giving consideration to the density of the transported material ρ_s [kg/m^3], the volumetric flow rate of solid particles Q_{sm} [Mg/h] may be determined, which provides information on the hourly use of the backfill materials, and thus the necessary efficiency of the dosing equipment [7].

In case of fine-grain mixtures, which permanently bond a certain amount of water both physically and chemically, the efficiency of filling the voids is difficult to determine. It may be generally assumed that the efficiency of filling the voids corresponds to the volumetric flow rate of the mixture, decreased by the amount of water which – in given conditions – is neither bonded during the solidification process nor absorbed by the surrounding gobs.

3. The characteristics and assessment of geometric parameters of a gravitational pipeline system

A diagram of an example of a pipeline route serving the hydraulic transport of mixtures has been presented in figure 2, while the basic geometrical parameters of the pipeline course in concern have been presented in table 1. Based on the above, a hydraulic profile of the pipeline course has been developed and presented in figure 3.

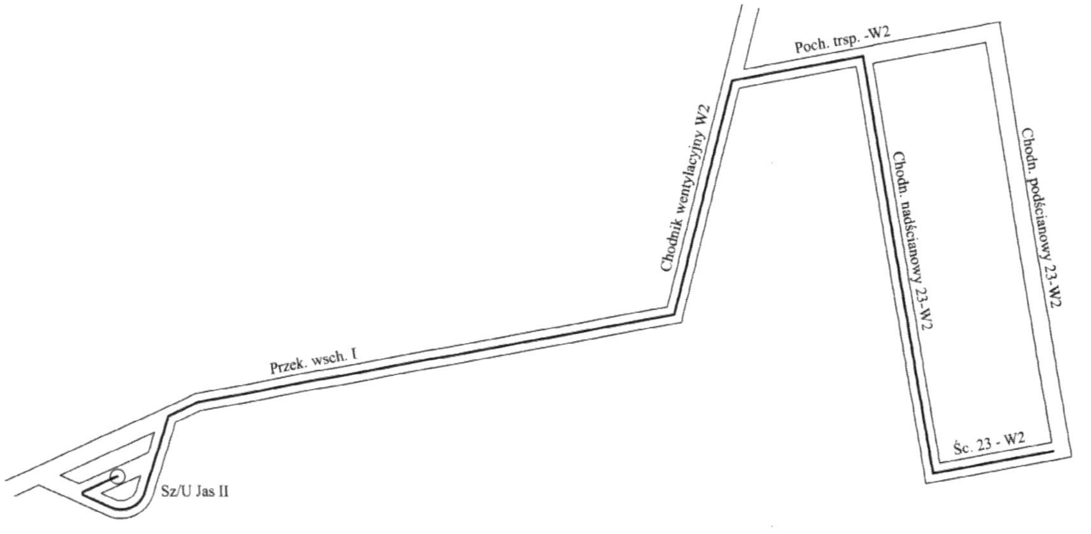

Figure 2. Exemplary route of pipeline for gravitational hydraulic transport of fine-grained mixture.

Table 1. General geometric parameters of transportation pipeline.

No.	Pipeline route (name of working)	Length of pipeline segment (m)	Height of begin/end of pipeline segment		Height difference of pipeline segment ΔH (m)	Diameter of pipeline segment D_i (m)	Equivalent length of pipeline segment L_{ekw} (m)
			H_p (m)	H_k (m)			
1	Szyb Jas II	660.0	258.1	-401.9	660.0	0.150	51.3
2	Podszybie poz. 400	56.0	-401.9	-401.9	0.0	0.090	56.0
3	Przekop Wschodni I	1718.0	-401.9	-395.8	-6.1	0.090	1718.0
4	Skrzyżowanie	30.0	-395.8	-395.7	-0.1	0.090	30.0
5	Przekop oddziałowy południowy	388.0	-395.7	-393.9	-1.8	0.090	388.0
6	Pochylnia wentylacyjna W2	366.2	-393.9	-449.3	55.4	0.090	366.2
7	Chodnik wentylacyjny W2 w pokł. 510/1-2	465.4	-449.3	-484.9	35.6	0.090	465.4
8	Pochylnia transportowa W2	154.8	-484.9	-523.3	38.4	0.090	154.8
9	Chodnik nadścianowy 23-W2	744.2	-523.3	-619.5	96.2	0.090	744.2
10	Ściana 23-W2	112.0	-619.5	-640.7	21.2	0.090	112.0
11	Sum	**4694.7**			**898.8**		**4086.0**

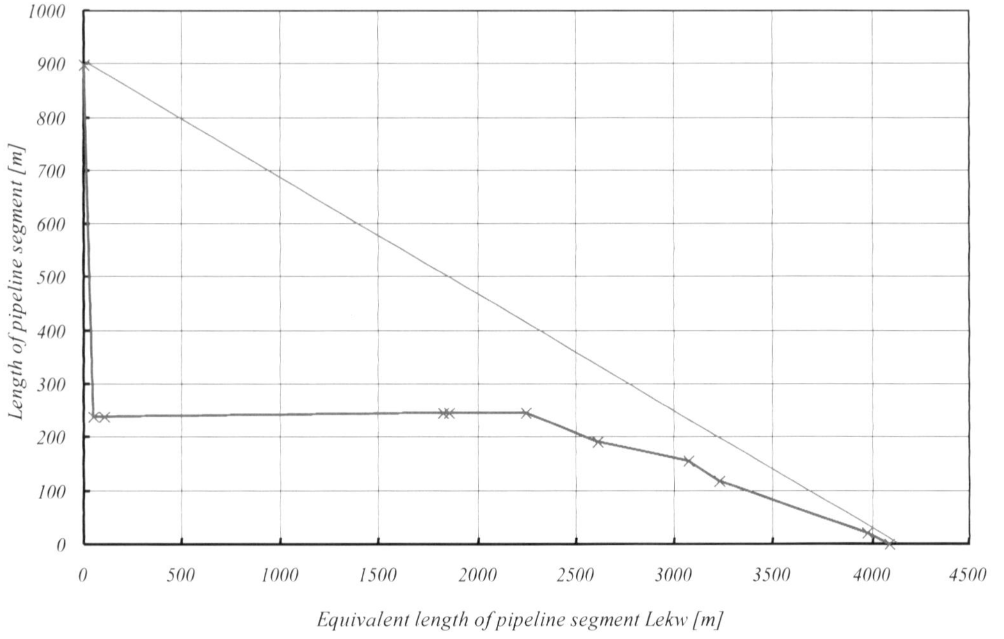

Figure 3. Hydraulic profile of pipeline installation for transport of fine-grained mixtures.

It follows from the table 1 that values of general geometric parameters of pipeline installation amount:

- Total length of pipeline.
- Total length of segments of 0,150 m diameter.
- Total length of segments of 0,09 m diameter.
- Equivalent diameter of the pipeline.

- Height difference between inlet and outlet of the pipeline.
- Equivalent length to height difference ratio.

The considered course of the pipeline serving the transport of fine-grain mixtures is characterized by a correct hydraulic profile that ensures favourable flow conditions. Considering the ratio of the transport length and the difference between the altitudes of the inlet and the outlet, one may initially conclude that it should ensure safe and effective flow in the gravitational hydraulic transport system.

4. Methods and results of tests of rheological parameters of fine-grain hydromixtures

Depending on the requirements of technology and the transportation possibilities, mixtures with different concentration of solid parts are applied in mining. As a result, there is a high variability of the rheological parameters of water and ash mixtures. A change in these parameters has a direct impact on the future rates of the gravitational hydraulic transport. The paper exhibits the changes of flow parameters in a pipeline system depending on the changes of the rheological parameters of the hydromixture. Fine-grain materials in the form of power-production ash from a fluidized bed boiler and varying proportions of water and solid particles were used to produce the hydromixtures. In the mining industry it is assumed that table spread is the measure of consistency, which, in line with PN-G/11011:1993, is the diameter of the circle formed upon spilling the hydromixture from a Ford cup onto a horizontal glass surface [2].

The rheological properties of the mixtures used in gob grouting were determined by means of rotational viscometry using the Bingham's rheological model to describe the rheological properties of fine-grain mixtures. The determined values have been presented in table 2.

Table 2. General rheological properties of fly ash – water mixture.

No.	Table spread R (mm)	Density of a mixture ρ_m (kg/m^3)	Volume concentration of a mixture C_V (-)	Dynamic viscosity η_B (Pa·s)	Yield point τ_0 (Pa)
1	155	1481	0.3195	0.223	17.79
2	160	1484	0.3183	0.201	14.90
3	170	1486	0.3171	0.173	10.65
4	180	1489	0.3160	0.146	7.55
5	200	1445	0.2958	0.111	4.06
6	205	1448	0.2947	0.103	3.48
7	210	1450	0.2936	0.096	2.97
8	215	1452	0.2924	0.09	2.45
9	235	1415	0.2754	0.071	1.2
10	245	1417	0.2743	0.065	0.84
11	250	1419	0.2732	0.061	0.7
12	255	1421	0.2722	0.058	0.55

As it can be concluded from rheological properties measurements, which results have been presented in table 2, that within the range of mixture table spread from 155 m to 255 m, their density varies from 1481 kg/m^3 to 1421 kg/m^3. Increasing table spread of mixtures is accompanying by decrease of dynamic viscosity in the range from 0,223 Pa·s to 0.058 Pa·s. Similar reaction occurs in case of yield point, which decreases from 17.29 Pa to 0.55 Pa by increasing table spread. It can be assumed that increase of table spread from 155 m up to 255 m results in reduction of dynamic viscosity by an order of magnitude and by two orders of magnitude for tables spread.

5. Determination and analysis of dependencies between the parameters of hydraulic transport of water and ash mixtures and their rheological parameters

The parameters of hydraulic transport parameters of water and ash mixtures defined in table 2 have been determined in line with the methods presented in section 2 of the paper and have been exhibited in table 3.

Table 3. Parameters of hydraulic transport of mixtures in an exemplary transport pipeline.

No.	Table spread R (mm)	Flow velocity v_m (m/s)	Volumetric flow rate (m^3/godz) mixture Q_m	solid parts Q_s	Water Q_w	Mass transport rate of solids Q_{sM} (Mg/godz)	Fell efficiency Q_p (m^3/godz)
1	155	2.430*)	55.65	17.78	37.87	44.56	23.11
2	160	2.921*)	66.90	21.29	45.60	53.65	27.68
3	170	2.426	55.56	17.62	37.94	44.64	22.91
4	180	2.558	58.58	18.51	40.07	47.14	24.06
5	200	2.743	62.82	18.58	44.24	46.57	24.16
6	205	2.795	64.01	18.86	45.15	47.52	24.52
7	210	2.845	65.16	19.13	46.03	48.45	24.87
8	215	2.892	66.23	19.37	46.86	49.33	25.18
9	235	3.078	70.49	19.41	51.08	48.65	25.24
10	245	3.099	70.97	19.47	51.50	49.05	25.31
11	250	3.140	71.91	19.65	52.26	49.77	25.54
12	255	3.174	72.69	19.78	52.91	50.39	25.72

*) laminar flow

For each of the mixtures under consideration, the following flow parameters have been determined:

- flow speed of the mixture in a 0.09 diameter pipeline, v_m,
- volumetric flow rate of the mixture in the pipeline, Q_m:
- volumetric stream of solid particles in the mixture Q_s,
- volumetric stream of water in the mixture Q_w,
- mass flow of solid particles in the mixture Q_{sM},
- efficiency of backfilling (volume of solid particles) Q_p.

The flow rate of water and solid particles, converted into mass units, constitutes a guideline for the fine-grain mixtures production plant to achieve a proper mixture production rate and a proper ratio of water and solid particles. The determined numerical values reflect the potential transport capability of the system, depending on the applied recipe of the mixture. To achieve the capabilities, the hydraulic gravitational transport system must be fully supplied. In practice, the actual operating conditions of the analyzed fine-grain mixtures transport system are dependent on the efficiency parameters of the water and ash mixture production system. The determined flow parameters may be assumed to be optimal in view of the operating parameters of the mixer, the dosing equipment and the pumps in the backfilling mixtures production plant.

While analyzing the calculated results of flow of ash and water substances in the pipeline defined in section 3 of the paper that have been presented in table 5.1, it may be concluded that the system ensures advantageous flow conditions in a wide range of concentrations of solid particles. The flow rate of fine-grain mixtures characterized by fluidity in the range from 155 mm to 255 mm is from approximately 55.56 m^3/h to 72.69 m^3/h. In the fluidity range in concern, no significant differences of efficiency occur at a varying fluidity. At a fluidity of the mixture in the range from 155 mm up to 255 mm, the change of the flow speed changes from 2.43 m/s to 3.17 m/s while the mass flow rate of the solid particles changes from approximately 37.9 m^3/h to 50.4 m^3/h. The approximate efficiency of filling the voids (Q_p) will assume the value from 22.9 m^3/h to 25.7 m^3/h, depending on the fluidity of the mixture.

The choice of the fluidity of the fine-grain mixture (concentration of the mixture) has a low impact on the achieved efficiency of void filling and the use of the industrial wastes for the production of the mixture. The amount of water required for the production, however, increases along with the increase of the fluidity – and the migration properties of the mixture are changing as well. From the point of view of fire prevention using the above mixtures, it should be underlined that it is disadvantageous to use both mixtures characterized by excessive density (low fluidity range, quick loss of absorbability) as well as ones that are too thin (flooding the gobs with water without proper filling which is supposed to provide a durable insulation).

6. Summary and conclusions

The paper presents tests of rheological properties of fine-grain hydromixtures produced based on power-production ashes and presents their impact on the flow parameters in a gravitational pipeline. The conducted tests and calculations allowed to formulate the following conclusions:

- The considered path of the pipeline for the transport of fine-grain mixtures ensures correct geometric parameters allowing for an effective gravitational transport of fine-grain mixtures.
- Based on the tests of rheological parameters of water and ash hydromixtures, it may be asserted that along with the increase of fluidity in the range from 155 mm to 255 mm, the value of the dynamic viscosity decreases in the range from 0.223 Pa·s to 0.058 Pa·s, the flow threshold decreases in the range from 17.79 Pa to 0.55Pa, while the density decreases in the range from 1481 g/dm^3 to 1421 g/dm^3.
- The analysis of flow parameters of water and ash mixtures based on their rheological parameters, conducted in section 5, allows to conclude that the volumetric flow rate of the hudromixture changes along with the increase of the fluidity (from 155 to 255mm) in the range from 55.65 m^3/h to 72.69 m^3/h, while the efficiency of backfilling increases in the range from 23.11 m^3/h to 25.72 m^3/h.
- Forecasts of the operating parameters of the system for the transport of ash and water mixtures based on laboratory tests of the rheological parameters of mixtures characterized by variable fluidity, have exhibited that the system provides the possibility of effective hydraulic transport of the backfilling mixtures within the entire range of the fluidity, that is from 155 mm to 255 mm.
- It should be noted that different types of fly ashes used for the production of water-ash mixtures may exhibit properties that are very different from those presented in the paper. In case of using other types of materials to produce the mixtures, the tests of their rheological properties must be conducted each time and recipes ensuring the effective hydraulic transport and the fulfilment of the technology requirements must be developed.
- In case of the determined hydraulic transport parameters, the volumetric flow rate of the mixture will be within the range from 55 m^3/h to 72 m^3/h. To produce such an amount of mixture, it is necessary for the plant producing the mixtures to mix fine-grain fly ashes in the maximal amount up to 50 Mg/h at a flow rate of water of up to approximately 53 m^3/h.

References

[1] Palarski J 1998 Design and use of Backfill for Support in Polish Coal Mines *6th International Symposium on Mining with Backfill 14 – 16 April, Brisbane, Australia*
[2] Popczyk M 2002 Ekologiczne i górnicze aspekty przemysłowego wykorzystania odpadów energetycznych i poflotacyjnych w podziemiach kopalń *Thesis Wydział Górnictwa i Geologii Gliwice*
[3] Plewa F and Mysłek Z 2001 *Zagospodarowanie odpadów przemysłowych w podziemnych technologiach górniczych* (Gliwice: Wydawnictwo Politechniki Śląskiej)
[4] Palarski J 1997 Fill transportation and Sedimentation Mechanisms For Stope Filling *9th International Conference on Transport and Sedimentation Of Solid Particles, 2-5 Sept, Kraków, Poland*
[5] Palarski J 1981 *Hydrotransport* (Warszawa: WNT)
[6] Plewa F, Mysłek Z and Strozik G 1997 Parametry hydraulicznego doszczelniania zrobów zawałowych odpadami drobnofrakcyjnymi *III Szkoła Geomechaniki Ustroń Poland*
[7] Janoszek T, Lubosik Z, Łączny J and Prusek S 2014 Modelowanie gęstych zawiesin popiołów lotnych w rurociągu podsadzkowym *Przegląd Górniczy* **4** pp 90-8

Hazards and occupational risk in hard coal mines – a critical analysis of legal requirements

Marcin Krause

Faculty of Mining and Geology, Silesian University of Technology, Akademicka 2, 44-100 Gliwice, Poland

marcin.krause@polsl.pl

Abstract. This publication concerns the problems of occupational safety and health in hard coal mines, the basic elements of which are the mining hazards and the occupational risk. The work includes a comparative analysis of selected provisions of general and industry-specific law regarding the analysis of hazards and occupational risk assessment. Based on a critical analysis of legal requirements, basic assumptions regarding the practical guidelines for occupational risk assessment in underground coal mines have been proposed.

1. Introduction

The current condition of occupational safety and health in hard coal mining, and especially the numbers and indicators characterizing occupational accidents and diseases, are a reason for seeking modern strategies and tools that would protect the life and health of the employees more efficiently. The obligations related to the occupational risk were supposed to be one of the tools used to improve the occupational safety and health in mines. These obligations include, among other things, the assessment of occupational risk related to the performed work, taking up actions which decrease the occupational risk, documentation of occupational risk assessment results and informing the employees on the occupational risk and the rules of protection against hazards.

The notion of occupational risk may be defined as the possibility of occurrence of the results of hazards in the working environment, manifested as injuries, health deterioration or reduced physical and mental capabilities. Generally, the term may be interpreted as the risk related to the performed profession. In more detail, it would be defined as the risk related to a given workstation or, to be precise, the risk related to the performance of work.

The form of implementation of legal requirements related to occupational hazards and risk is specified by the provisions of general and industry-specific law, including, among others: the requirement to keep a documentation of the occupational risk and the applied necessary preventive measures as well as the development and updating of the safety and health document for the employees of the mining plant. Although these obligations are formally met, the practical aspect still raises many doubts – among other things – in terms of the classification of the mining hazards (including natural hazards) and the rules of selection of occupational risk assessment methods adequate to the specific characteristics of the hazards in hard coal mines.

A detailed analysis of the general and industry-specific law regarding the occupational safety and health in hard coal mines has been presented, among other sources, in publications [1, 2]. The general guidelines for occupational risk assessment are given by the Polish Standard PN-N-18002 [3]. The basics of the methodology of occupational risk assessment included, among others in CIOP-PIB publications [4, 5] and in the author's publications [1, 2, 6, 7, 8].

2. Analysis of selected provisions of the general law

The problems related to the assessment of the occupational risk started to gain increasing importance in the legislation of the EU countries since 1989, after the adoption of the 89/391/EEC directive regarding the introduction of measures to encourage improvements in the safety and health of workers at work [9], hereinafter called the 'framework directive' due to its leading nature in regard of detailed directives.

The Directive [9] contains the basic legal requirements regarding the obligations of the employers and employees related to the protection of life and health in the work environment, as well as it serves as the legal basis for the individual directives in terms of the provisions regarding the individual sections of the national economy, special risk groups, types of work processes and types of hazards.

The Framework Directive constitutes a new approach to the problems of protection of health and life in work environment, including: active prevention of hazards, cooperation of employees in management, ensuring continuous improvement of the health and safety conditions at work and the integration of various health and safety-related activities.

The Directive [9] has introduced new obligations for the employers related to the health and safety of the employees, among other things related to the prevention of risks, such as:

- Art. 6. 1. Within the context of his responsibilities, the employer shall take the measures necessary for the safety and health protection of workers, including the prevention of occupational risks and provision of information and training, as well as provision of the necessary organization and means. The employer shall be alert to the need to adjust these measures to take account of changing circumstances and aim to improve existing situations.
- Art. 6. 2. The employer shall implement the measures referred to in the first subparagraph of paragraph 1 on the basis of the following general principles of prevention: (a) avoiding risks; (b) evaluating the risks which cannot be avoided; (c) combating the risks at source; ... (i) giving appropriate instructions to the workers.
- Art. 6. 3. Without prejudice to the other provisions of this Directive, the employer shall, taking into account the nature of the activities of the enterprise and/or establishment: (a) evaluate the risks to the safety and health of workers, inter alia in the choice of work equipment, the chemical substances or preparations used, and the fitting-out of workplaces …

The provisions of the Framework Directive have been introduced to the Polish Law starting with the Labour Code Act [10] and ending with numerous executive acts.

The highest act in Poland is the Constitution of the Republic of Poland [11], which, in art. 66. 1 provides that everyone shall have the right to safe and hygienic conditions of work.

The legal basis regarding the occupational safety is constituted by the Act [10], which, in art. 15 states that employers are obliged to ensure healthy and safe working conditions for their employees.

The general rules regarding the organization of the occupational safety and health in workplaces are included in the chapter X of the Act [10] and in executive acts including the ordinance regarding special occupational safety and health provisions [12], hereinafter called the "general OSH provisions" due to their general application to all types of activities.

In line with the provisions of the Act [10]:

- Art. 207. § 1. The employer is responsible for health and safety in the work establishment.
- Art. 207. § 2. The employer is obliged to protect the health and life of employees by ensuring conditions of health and safety at work by the appropriate use of the achievements of science and technology.
- Art. 226. The employer must: 1) evaluate and provide documentation on any occupational risk connected with the work performed, and must apply necessary preventative measures decreasing the risk, 2) inform employees about any occupational risk related to the work performed, as well as on the principles of protection against dangers.

In line with the requirements of the ordinance [12]:

- § 40. 1. The employer shall provide systematic checks of the state of occupational safety and health with particular reference to the organisation of work processes, the technical condition of machinery and other technical equipment, and to determine methods to register irregularities and to remove them.
- § 39. 1. The employer shall implement the obligation to provide employees with safety and health at work, in particular by preventing the risks associated with their work, the Organization of work, the use of the necessary preventive measures and informing and training workers.
- § 39. 2. The obligation referred to in section 1, should be implemented on the basis of the General rules for prevention of accidents and work-related diseases, in particular by: 1) prevention of hazards; 2) carrying out the risk assessment of hazards, which may not be excluded; 3) removing hazards at the source of their formation; … 8) instructing employees on health and safety at work.
- § 39a. 1. The employer shall assess the occupational risk occurring while executing the work, in particular in the selection of equipment of workstations and workplaces, used chemical and biological substances and preparations, carcinogenic or mutagenic, and change the work organization. During the occupational risk assessment, all the factors in the working environment occurring while executing the work, and methods of performing the work shall be taken into account.
- § 39a.3 The employer shall keep a documentation of the occupational risk assessment and the applied necessary preventive measures. The document confirming the performance of the occupational risk assessment should give consideration to: 1) the description of the workstation under assessment, including a list of: a) the applied machinery, tools and materials, b) the performed tasks, c) the occurring at workstation hazardous, harmful and burdensome factors of the working environment, d) the applied individual and collective protection measures, e) the persons working at this workstation; 2) the results of the conducted occupational risk assessment per each of the work environment's factors and the necessary preventive measures to reduce the risk; 3) the date of the conducted assessment and the person performing the assessment.

The provisions of the Act [10] and the ordinance [12] constitute the general law regarding the occupational safety and health while the specific law includes, among other things:

- selected industries, such as mining, construction business, health care;
- selected types of work processes, such as using machines, manual handling;
- selected types of hazards, such as physical, chemical, biological and psychophysical;
- selected special risk groups e.g. juveniles, women including pregnant women.

Among many legal acts regarding the occupational safety and health that touch upon the hazards analysis and occupational risk assessment, the following may be listed:

- provisions regarding factors harmful to health [13, 14],
- provisions regarding the operation of machines [15],
- provisions regarding manual handling [16],
- provisions regarding the hazards related to chemical factors [17, 18],
- provisions regarding the exposure to harmful biological factors [19, 20],
- provisions regarding the exposure to noise and mechanical vibrations [21],
- provisions regarding the exposure to optical radiation [22],
- provisions regarding the exposure to ionizing radiation [23, 24, 25].

The Ordinance [13] specifies the following types of harmful conditions which require tests and measurements: chemical factors and dusts, carcinogenic and mutagenic factors, optical radiation (laser and non-laser), electromagnetic fields and radiation, cold and hot microclimate, noise and ultrasonic noise, mechanical vibrations (general and local). The rules for the assessment of ionizing radiation are subject to separate provisions [23, 25].

The ordinance [14] specifies the following types of harmful conditions, while providing the maximum allowable concentrations and intensities: chemical factors, dusts, cold and hot microclimate, noise and ultrasonic noise, mechanical vibrations (general and local), optical radiation (laser and non-laser), electromagnetic fields and radiation. The threshold dosages of ionizing radiation are included in the provisions of the ordinance [24].

3. The analysis of selected provisions of the industry-specific law

The problems of the assessment of occupational risk in underground mining started to gain increasing importance in the legislation of the EU countries since 1992, after the adoption of the 92/104/EEC directive regarding the minimum requirements for improving the safety and health protection of workers in surface and underground mineral-extracting industries [26].

The directive [26] imposed new responsibilities on the employers, related to the safety and health protection of the employees, related among other things to the development and updating the document regarding health and safety protection and the analysis of risk related to occupational safety and health:

- Art. 3. 2. The employer shall ensure that a document concerning safety and health, hereinafter referred to as "safety and health document", covering the relevant requirements laid down in Articles 6, 9 and 10 of Directive [9]. The safety and health document shall demonstrate in particular that: the risks to which workers at the workplace are exposed have been determined and assessed, adequate measures will be taken to attain the aims of this Directive, the design, use and maintenance of the workplace and of the equipment are safe.
- Art. 8. 1. To ensure that workers receive health surveillance appropriate to the health and safety risks they incur at work, measures shall be introduced in accordance with national law and/or practices.

The provisions of the directive [26] have been introduced to the Polish Law starting with the Geological and Mining Act [27] and ending with numerous executive acts.

In line with the provisions of the Act [27]:

- Art. 108. 2. The mining plant operations plan specifies: ... 2) specific activities necessary to ensure: ... d) the safety of persons residing in the mining plant, in particular concerning occupational safety and health.
- Art. 117. The entrepreneur is obliged to: 1) identify hazards associated with mining plant operations and try to implement the measures to prevent and remove these hazards; ... 4) evaluate and document the occupational risk and apply the necessary solutions that mitigate this risk, including the drafting of the document of the safety and health protection ...

In the range of mining, the main executive acts of the Act [27] are, among others, the provisions related to the operation of the plant, the natural hazards, mining rescue operations and qualifications necessary in mining and mining rescue operations.

In terms of occupational safety and health, since 01.07.2017, the basic executive act of the Act [27] are the provisions regarding the operation of underground mining plants [28] (including: section II 'Occupational safety and health and fire safety', section II 'Ventilation', section IV 'Working conditions and health protection', section V 'Hazards occurring in the operation of a mining plant').

Section II of the ordinance [28], among other things specifies the detailed rules for the assessment and documentation of occupational risk and for the application of necessary preventive measures decreasing the risk, in the form of an employees' safety and health protection document.

In line with the requirements of the ordinance [28]:

- § 5. 1. Analyses and tests required for the safe operation of a mining plant, including the assessment and documentation of occupational risk and the application of necessary solutions for decreasing the risk, shall be conducted by the entrepreneur.
- § 6. 1. Before commencing the works, the entrepreneur shall draft an employees' safety and health protection document, hereinafter called the 'safety document'.
- § 6. 4. The mining plant's operation manager shall inform the employees of the mining plant the employees of the entities referred to in art. 121. 1 of the Act [27] on the contents of the safety

document or a relative part thereof, especially concerning the occupational risk assessment and the application of the necessary solutions applied to decrease the risk.

- § 218. 1. Tests and measurements of harmful factors and ionizing radiation in the working environment shall be conducted in the mining plant – in a manner specified by the provisions issued based on the art. 228 § 3 of the Act dated June 26th, 1974 – Labour Code [10], and based on the art. 25. 1 of the Act dated November 29th 2000 – Nuclear Law [23], with the exclusion of dusts, climatic hazards and harmful factors specified in § 142. 2, for which the tests are conducted in line with the provisions of the ordinance.
- The provisions of the ordinance [28] contain a reference to the following types of hazards:
- mining hazards: rock burst hazard, methane hazard, gas and rock breakout hazard, coal dust explosion hazard, climatic hazard, water hazard, natural radioactive substances hazard, fire hazard (section V, chapters 1-9),
- harmful gases: carbon dioxide, carbon oxide, nitrogen oxide, sulphur dioxide, hydrogen sulphide (section IV, chapter 1),
- harmful dusts (section IV, chapter 2),
- sources of ionizing radiation (section VI, chapter 8).

The basic hazards occurring in mining plants include natural hazards, for which the following provisions constitute the legal basis:

- until 30.06.2017, the provisions of the ordinance [29] encompassing: rock burst hazard, methane hazard, gas and rock breakout hazard, coal dust explosion hazard, water hazard, eruption hazard, hydrogen sulphide hazard, natural radioactive substances hazard, harmful dusts activity hazard;
- from 01.07.2017, the provisions of the ordinance [30] encompassing: rock burst hazard, methane hazard, gas and rock breakout hazard, coal dust explosion hazard, climatic hazard, water hazard, rockslide hazard, eruption hazard, hydrogen sulphide hazard, radioactive substances hazard (radiation hazard).

4. Summary and conclusions

Based on the critical analysis of selected provisions of the general and industry-specific law and the practical functioning thereof in selected hard coal mines, the following conclusions have been formulated:

- The current provisions are unsatisfactory. This concerns, among other things, the inconsistencies between the general and industry-specific law and the lack of practical guidelines for the assessment of occupational risk.
- Differing definitions of the occupational risk are an inconsistency. It is often referred to the workstation (ordinance [12]), while it should be related to the conducted work (act [10]).
- The lack of practical rules for the selection of the occupational risk assessment method both in the legal acts and in the PN-N-18002:2011 [3] standard is a basic problem.
- In terms of the occupational risk assessment in hard coal mines, a division of workstations into stationary and non-stationary may be applied, not only giving consideration to the fixed or varying location of workplace but also to the criteria of variability of the types of the work processes and the working conditions.
- Another important matter is constituted by the lack of coherent rules for the classification of hazards in mining plants. This results both from the analysis of legal acts and the industry-specific literature. It may be exemplified by the classification of hazards depending on the energy location criteria (natural, technical, personal) that is often applied in mines.
- As far as the assessment of occupational risk in hard coal mines is concerned, a classification of hazards depending on the character/type of activity (physical, chemical, biological, psychophysical) or the criterion of consequence/results of the activity (hazardous, harmful, burdensome) may be applied.

- In terms of the mining hazards, many controversies are caused by both the currently binding [30] and the out-dated [29] classification of natural hazards. The controversies are raised, among other things, by endogenous fires hazard, dust hazard and ionizing radiation hazard.
- The specific character of the working environment and hazards occurring in hard coal mines constitutes a difficulty in terms of the duty to assess and document the occupational risk, especially in the range of the variability of the workplace, the work process and the working conditions.
- Despite the advancement in the methodology of occupational risk assessment in mining, the general rules of developing and conducting the risk assessment and the qualitative method of risk estimation (risk matrix) in line with the PN-N-18002 [3] standard, still remain dominant.
- A shift from the traditional approach to documentation of the occupational risk assessment for a given workstations (stationary workstations) towards the assessment of the risk related to the conducted work (non-stationary workstations) should be made. This should be done while keeping in mind the criteria of the location of the workplace, the type of work process and the working conditions.
- The approach to informing the employees on the occupational risk should be altered from a complex information regarding the results of the risk assessment per a given workstation (multi-page document exhibited during OHS trainings) to the presentation of the most important information regarding the hazards and the occupational risk related to the conducted work (1 or 2 page document exhibited while changing the workplace, the work process or the working conditions).

References

[1] Krause M and Romanowska-Słomka 2014 *Podstawy bezpieczeństwa i higieny pracy* [Basics of occupational safety and health] (Wałbrzych: Państwowa Wyższa Szkoła Zawodowa w Wałbrzychu)

[2] Krause M 2012 *Zasady doboru metod oceny ryzyka zawodowego w aspekcie zróżnicowanego oddziaływania niebezpiecznych i szkodliwych czynników środowiska pracy w kopalniach węgla kamiennego* [Selection principles of the occupational risk assessment methods in view of variable impact of hazardous and harmful factors of working environment in hard coal mines] (Gliwice: Wydawnictwo Politechniki Śląskiej)

[3] Polska Norma PN-N-18002:2011 Systemy zarządzania bezpieczeństwem i higieną pracy. Ogólne wytyczne oceny ryzyka zawodowego [Occupational safety and health management systems. General guidelines for occupational risk assessment] (Warszawa: Polski Komitet Normalizacyjny)

[4] Koradecka D (ed.) 2008 *Bezpieczeństwo i higiena pracy* [Occupational safety and health] (Warszawa: CIOP-PIB)

[5] Zawieska W (ed.) 2009 *Ryzyko zawodowe. Metodyczne podstawy oceny* [Occupational risk. Methodological basis for assessment] (Warszawa: CIOP-PIB)

[6] Krause M 2016 Ocena ryzyka zawodowego – wybrane problemy analizy ryzyka [Occupational risk assessment – selected problems of risk analysis] *Wiadomości Górnicze* **5** pp 310-15

[7] Krause M 2017 Ocena ryzyka zawodowego w aspekcie miejsca i warunków wykonywania pracy [Occupational risk assessment in terms of place and conditions of work performance] *Zeszyty Naukowe Politechniki Śląskiej Seria Organizacja i Zarządzanie* Volume 100 pp 209-224

[8] Krause M 2016 Zarys metodyki oceny ryzyka zawodowego w aspekcie analizy metod badań [Outline of methodology of occupational risk assessment in terms of the analysis of research methods] *Zeszyty Naukowe Wyższej Szkoły Zarządzania Ochroną Pracy w Katowicach* Tome 12 **1** pp 74-88

[9] Council Directive No. 89/391/EEC dated June 12th, 1989, regarding the introduction of measures to encourage improvements in the safety and health of workers at work. OJ EC L 183 dated

29.06.1989; with amendments
[10] Act dated June 26th, 1974, Labour Code. 2016 Journal of Laws, pos. 1666; with amendments
[11] The Constitution of the Republic of Poland dated April 2nd, 1997, approved by the National Assembly on April 2nd 1997, approved by the Nation in a constitution referendum on May 25th 1995, signed by the President of the Republic of Poland on July 16th, 1997. 1997 Journal of Laws No. 78, pos. 483; with amendments
[12] Ordinance of the Minister of Labour and Social Policy dated September 26th, 1997, regarding the general provisions on the occupational safety and health. 2003 Journal of Laws No. 169, pos. 165; with amendments
[13] Ordinance of the Minister of Health dated February 2nd 2011, regarding the tests and measurements of harmful factors in work environment. 2011 Journal of Laws No. 33, pos. 166
[14] Ordinance of the Minister of Labour and Social Policy dated June 6th, 2014, regarding the highest allowable concentrations and intensities of factors harmful to health in working environment. 2014 Journal of Laws, pos. 817; with amendments
[15] Ordinance of the Minister of Economy and Labour dated August 5th, 2005, regarding the occupational safety and health during works causing an exposure to noise or mechanical vibrations. 2005 Journal of Laws No. 157, pos. 1318
[16] Ordinance of the Minister of Labour and Social Policy dated March 14th, 2000, regarding the occupational safety and health in manual handling. 2000 Journal of Laws No. 26, pos. 313; with amendments
[17] Act dated February 25th, 2011, on the chemical substances and their mixtures. 2015 Journal of Laws, pos. 1203; with amendments
[18] Ordinance of the Minister of Health dated December 30th, 2004, regarding the occupational safety and health in relation to the occurrence of chemical factors at the workplace. 2005 Journal of Laws No. 11, pos. 86; with amendments
[19] Act dated December 5th, 2008, on preventing infections and infectious diseases among people. 2016 Journal of Laws, pos. 1866
[20] Ordinance of the Minister of Health dated April 22nd, 2005, regarding harmful biological factors in work environment and the protection of health of the employees exposed to these factors. 2005 Journal of Laws No. 81, pos. 716; with amendments
[21] Ordinance of the Minister of Economy dated October 30th, 2002, regarding the minimal requirements concerning the occupational safety and health while using machines by employees. 2002 Journal of Laws No. 191, pos. 1596; with amendments
[22] Ordinance of the Minister of Labour and Social Policy dated May 27th, 2010, regarding the occupational safety and health in works related to exposure to optical radiation. 2013 Journal of Laws, pos. 1619
[23] Act dated November 29th, 2000, Nuclear Law. 2017 Journal of Laws, pos. 576
[24] Ordinance of the Council of Ministers dated January 18th, 2005, regarding the threshold dosages of ionizing radiation. 2005 Journal of Laws No. 20, pos. 168
[25] Ordinance of the Council of Ministers dated March 23rd, 2007, regarding the requirements of registration of individual dosages. 2007 Journal of Laws No. 131, pos. 913
[26] Council Directive No. 92/104/EEC dated December 3rd, 1992, regarding the minimum requirements for improving the safety and health protection of workers in surface and underground mineral-extracting industries. OJ EC L 404 dated 31.12.1992; with amendments
[27] Act dated June 9th, 2011, Geological and Mining Law. 2016 Journal of Laws, pos. 1131; with amendments
[28] Ordinance of the Minister of Energy dated November 23rd, 2016, regarding the detailed requirements for the operation of the underground mining plants. 2017 Journal of Laws, pos. 1118
[29] Ordinance of the Minister of Internal Affairs and Administration dated June 14th, 2002, regarding natural hazards in mining plants. 2002 Journal of Laws No. 94, pos. 841; with amendments

IOP Conf. Series: Materials Science and Engineering **268** (2017) 012013 doi:10.1088/1757-899X/268/1/012013

[30] Ordinance of the Minister of Environment dated January 29th, 2013, regarding natural hazards in mining plants. 2014 Journal of Laws, pos. 1129; with amendments

IOP Conf. Series: Materials Science and Engineering **268** (2017) 012014 doi:10.1088/1757-899X/268/1/012014

Motivation and proposal of important categories of the work of mine rescuer

Aneta Grodzicka[1], Jan Szlązak[1] and Piotr Buchwald[2]

[1]Faculty of Mining and Geology Silesian University of Technology, Akademicka 2, 44-100 Gliwice, Poland
[2]Central Mines Rescue Station in Bytom, Chorzowska 25, 41-902 Bytom

aneta.grodzicka@polsl.pl

Abstract. The paper presents the results of the conducted research using the questionnaire. Based on published categories of motives present in the work situation, 12 questionnaires were sent directly to the salvage miners' professional group. On the basis of the results obtained, 12 categories of importance of motives present in the life of the rescuer were proposed. After identifying the categories, a scheme of motives for the meaning of work in the life of the rescuer was proposed, which distinguishes six basic motives of the significance of the work of a lifeguard.

1. Introduction

Professional work is a very important element for the functioning of every adult. Man needs to use his ability to meet job requirements and experience interpersonal support, which is considered to be the main characteristics of work conditioning the physical and mental well-being of the individual [1]. The dynamics of changes in labor relations has been high and it is defined as follows [2]:

- The concept of "work" has been separated from the concept of 'employment' by explaining less and less work is realized through constant employment
- Employment requires the activity of an individual, for example, the nature
- Competitiveness through the flexibility of employment and thus the exchange of employees
- Employee exchange raises dissatisfaction received very negatively since it increases the sense of uncertainty,
- The permanent employment model from graduation to retirement disappears. There is a 'break' model in working life

According to T. Tomaszewski, the work is distinguished in six categories of motive groups present in the work situation [3]:

- Successes and failures - every person commits himself to perform certain tasks and is motivated by attaining the intended goal.
- Direct benefits and losses - any professional decision is dictated by the possibility of earning economic profits.
- Safety and Hazards - every person in the workplace strives to ensure a sense of security.
- Relationships between people - each person strives for the right and the same correct social relationship through kindness, support, acceptance.

- Attractiveness and aversiveness of particular activities - Any activity performed at work can be considered as or not liked by the unit performing the task.
- Comfort and discomfort at work - every man strives for comfortable working conditions to feel good in his workplace, while uncomfortable working conditions lead to an additional effort of the worker, where overcoming inner resistance leads him to perform work at the limits of his strength and capabilities.

Mine rescue is a voluntary service. Every miner can apply for permission to become a lifeguard. After proper testing and fulfillment of the conditions, the mine rescue. Rescue is a job and a challenge. The basic motivation is to save another human "human life" that is invaluable. Rescuers are proud to be able to serve. As a result, research into the importance of work in the life of the rescuer has been made. Based on the aforementioned motive groups, a questionnaire was created and 2 questions were assigned to each group. In total, 12 questions were included in the questionnaire and the research was carried out on a group of rescuers

2. Appropriate research

At the turn of 2016 and 2017, a survey was conducted on a group of mine rescue workers from one of the hard coal mines. The research sample consisted of 94 rescuers, which accounted for 50% of the salvage status of the rescuers during the survey. The research was done during the training along with the theoretical part. Everyone was informed about the research and its purpose. In addition, they received information about the anonymity of the survey and a thorough explanation of how to complete the questionnaire.

Respondents rescued were three age brackets between 21-30 years, 31-40 years and over 40 years. The most common compartment (31-40 years) counted 58 rescuers, which constituted 61.7%. It can therefore be stated that the surveyors who took part in the study were the so-called senior employees not belonging to the smallest or oldest compartment (table 1).

Table 1. Age brackets of rescuers participating in the study.

Age of Mine Rescuers	21-30 years	31-40 years	>40 years	Sum
number	24	58	12	94
%	25.5	61.7	12.8	100

When analyzing rescuers in terms of seniority in salvage, the noticeable prevalence of rescuers is in the range of 6-10 years. Then in two intervals, 1-5 years and 11-15 years, there were 17% of the whole population tested (table 2).

Table 2. Lengths of service of rescuers participating in the study.

Mine Rescuers work life	1-5 years	6-10 years	11-15 years	16-20 years	>20 years
number	16	44	16	10	8
%	17.0	46.8	17.0	10.6	8.6

Table 3. Declared responses in the survey.

Lp.	Question asked in survey sheet	Answer Yes		Answer No	
		number	%	number	%
1.	Does the rescue action motivate you to achieve the goal of: ending the action as soon as possible?	90	95.7	4	4.3
2.	During the rescue operation motivates you to reach the goal: how to get the victims out as soon as possible?	94	100.0	0	0
3.	Are you happy with your monthly salary?	71	75.5	23	24.5
4.	Do you have the need to improve your qualifications for promotion?	62	66.0	32	34.0
5.	Do you feel a great sense of security by doing health and safety activities?	89	95.0	5	5.0
6.	Do you think your workplace is safe without any additional risk reduction that may occur?	21	22.3	73	77.7
7.	Do you feel accepted by your team mates?	93	99.0	1	1.0
8.	Can you count on support from your colleagues and supervisors?	90	96.0	4	4.0
9.	Do you do something you do not like?	30	32.0	64	68.0
10.	Is your position is dominated by the activities you liked?	86	91.5	8	8.5
11.	Are you happy with your job?	87	92.6	7	7.4
12.	Do you feel that you do your work on the edge of endurance?	3	3.2	91	96.8

Figure 1 shows the results of the question 'Does the rescue operation motivate you to achieve the goal of: ending the action as soon as possible?' Rescue survivors found that 65.7% of rescue operations were their goal. Rescue is a voluntary service and every rescuer wants to achieve a career goal, which is to complete the rescue operation.

The question 'Does the rescue operation motivate you to achieve the goal of: getting the victims out as soon as possible?' Indicates that for the 100% of the rescuers rescuing the victims is the achievement of the goal. Let us not forget that miners rescuers always claim 'we go to live' (figure 2).

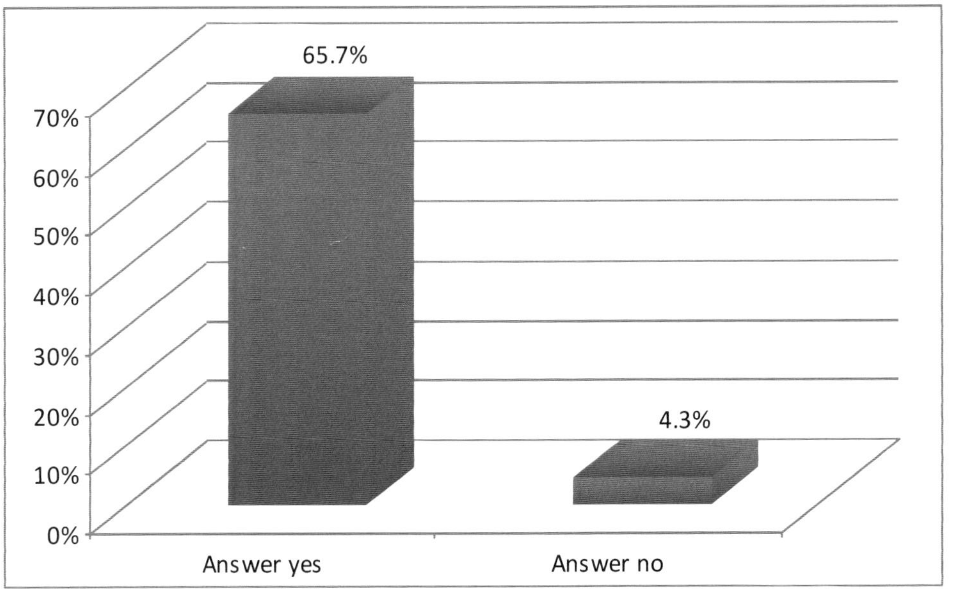

Figure 1. Declared answers to question No. 1.

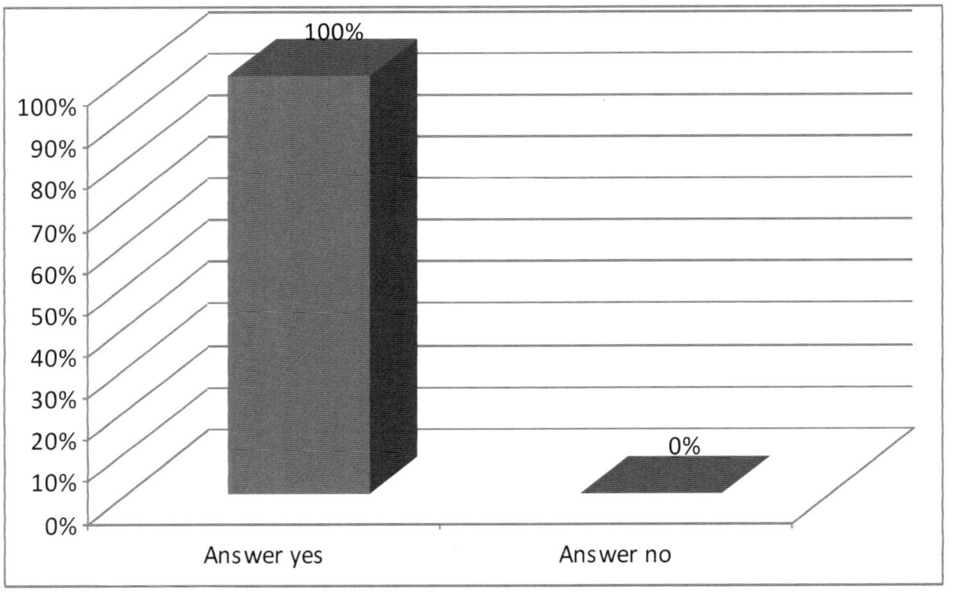

Figure 2. Declared answers to question No. 2.

Figure 3 shows the results of satisfaction with the monthly salary, 75.5% of the respondents confirmed satisfaction. We are now living in a time where young people are striving to rapidly raise their living conditions and consequently raise finances. And it was this group that most likely presented a negative opinion about satisfaction from a monthly salary.

'Do you have the need to improve your qualifications for promotion?' Question No. 4 in the questionnaire. Probably the young rescuers have declared 66% of the need to improve their qualifications. This is a positive trait that signifies a great commitment to the performance and seriousness of the work performed (figure 4).

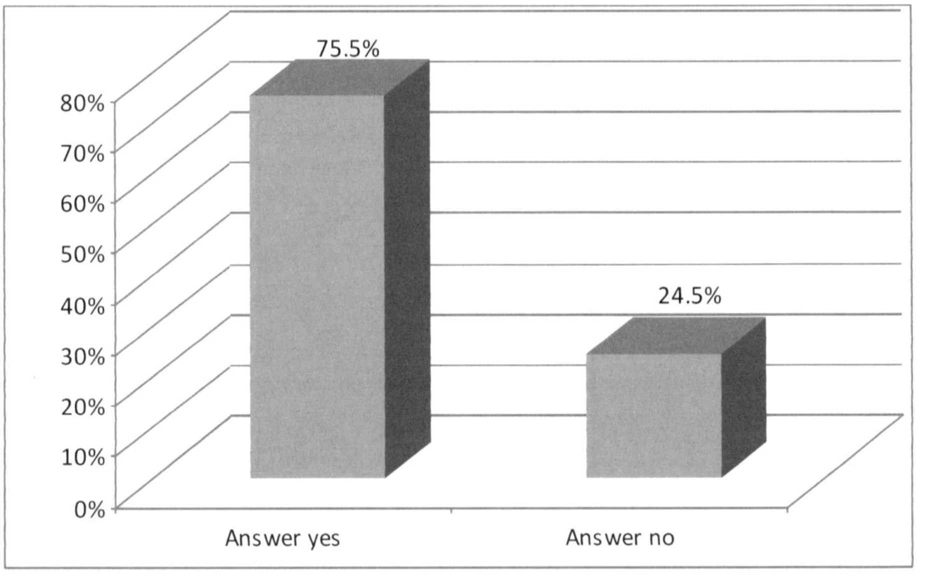

Figure 3. Declared answers to question No. 3.

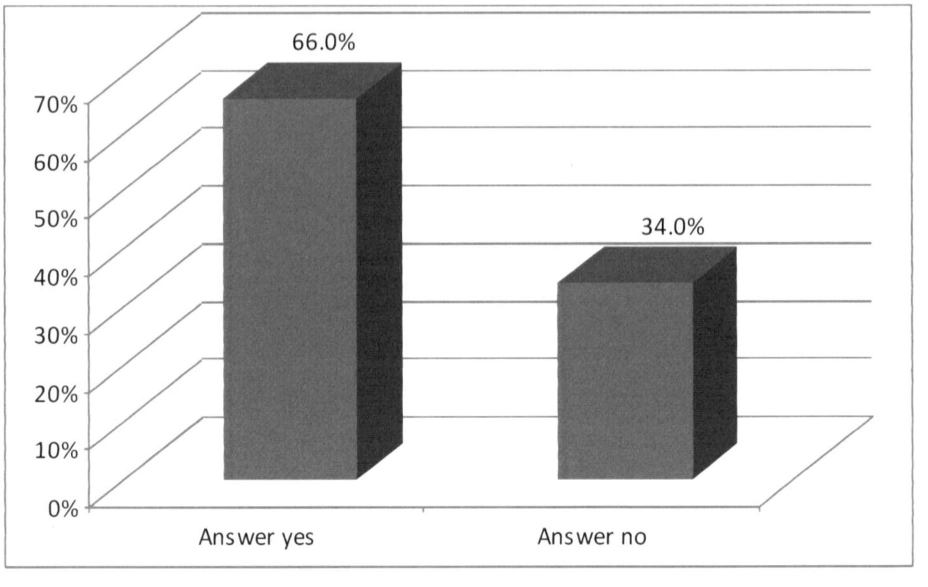

Figure 4. Declared answers to question No. 4.

Figure 5 shows the results of the questionnaire, 'Do you feel a great sense of security by doing your health and safety activities?' The 95% group confirmed that they are safe by doing their job in accordance with safety standards. Each rescuer has knowledge of the consequences of not following the principles of safe work, which may result in an accident at work.

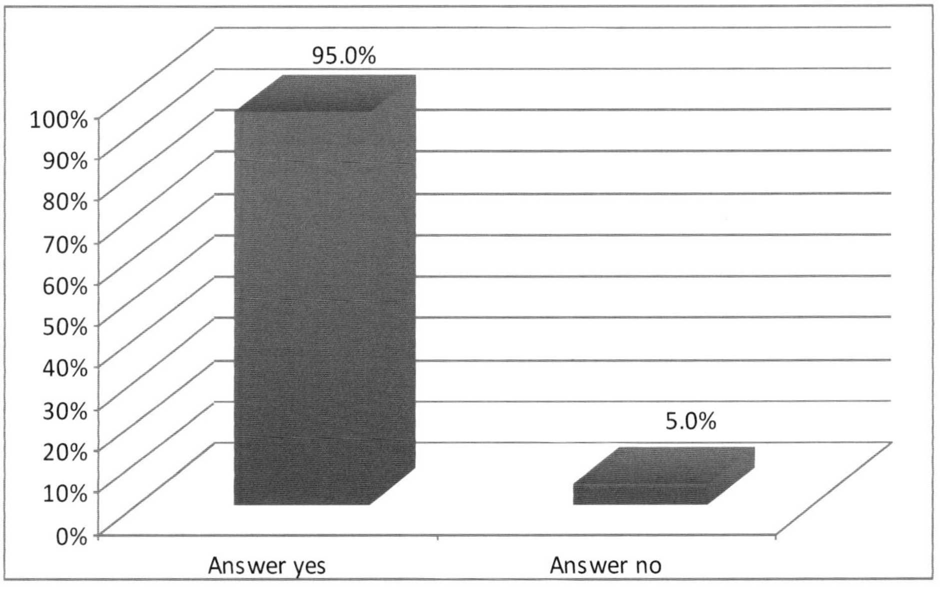

Figure 5. Declared answers to question No. 5.

Rescuers' response to 'Do you think your workplace is safe without any additional reduction of the risks that may occur?' Show rescuers are demanding a reduction in the risks that could occur in their workplace at 77.7%. Respondents are aware that any hazard can lead to even fatal consequences (figure 6).

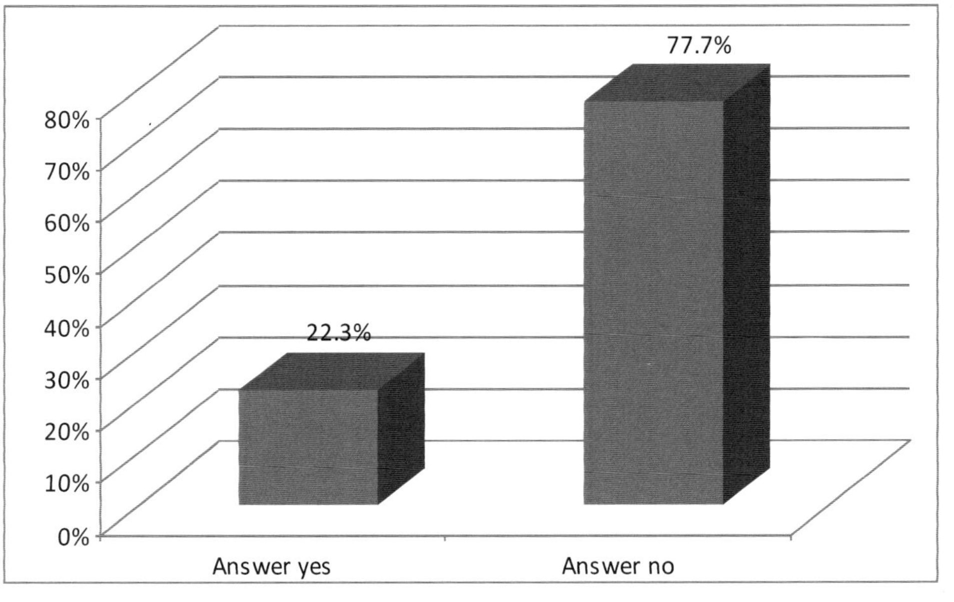

Figure 6. Declared answers to question No. 6.

Figure 7 shows the results of team acceptance by colleagues. Because rescuers work in teams, the results confirm the bond and acceptance of the individual units in the team. Only 1% of respondents expressed a disagreement, they could be people with small work placements, where team bond was not established or not yet participated in rescue operations.

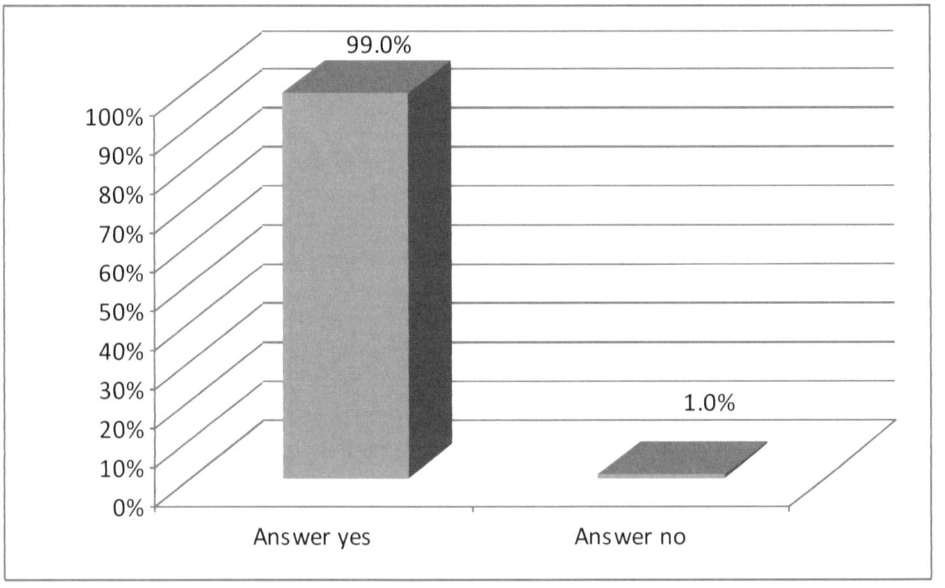

Figure 7. Declared answers to question No. 7.

Figure 8 shows the results of the question 'Can you count on support from your colleagues and supervisors?' The respondents in 96% rated the relationship with superiors and colleagues as very good, knowing they can count on them in every situation. This confirms the very good coordination and teamwork, which is undoubtedly one of the important characteristics of work in the rescue.

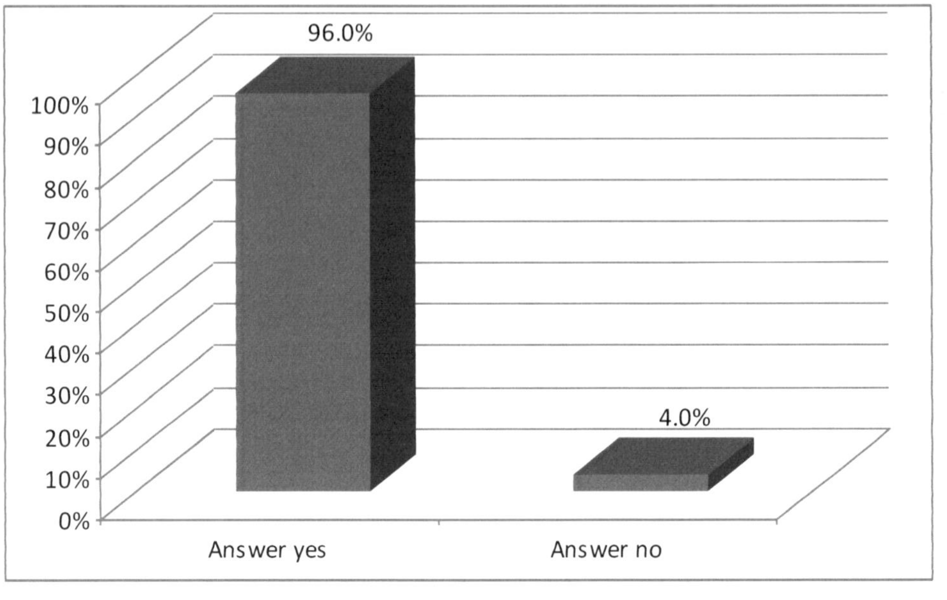

Figure 8. Declared answers to question No. 8.

Each of the employees, as well as the rescuers of the mining work, considers it more or less acceptable. This is not only a way of doing work, but also a number of training activities in the area of rescue operations. The surveyed respondents in 32% stated that their workstations performed work that they did not like. Most likely, this evaluation results from lack of experience.

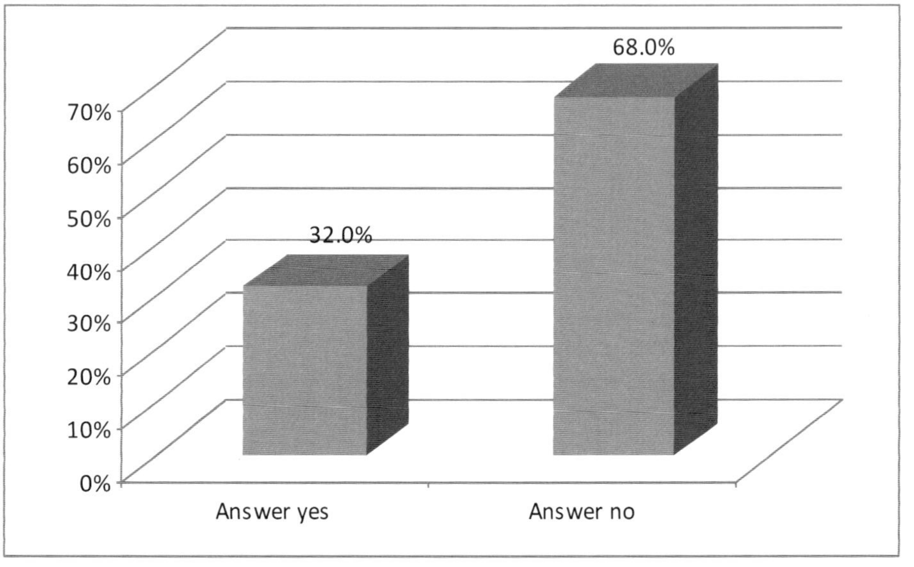

Figure 9. Declared answers to question No. 9.

Figure 10 shows the results of the rescue rescuers' declaration on the question, 'Do you like your favorite activities?'. According to 91.5% of respondents, they consider their work to be the most popular job they enjoy.

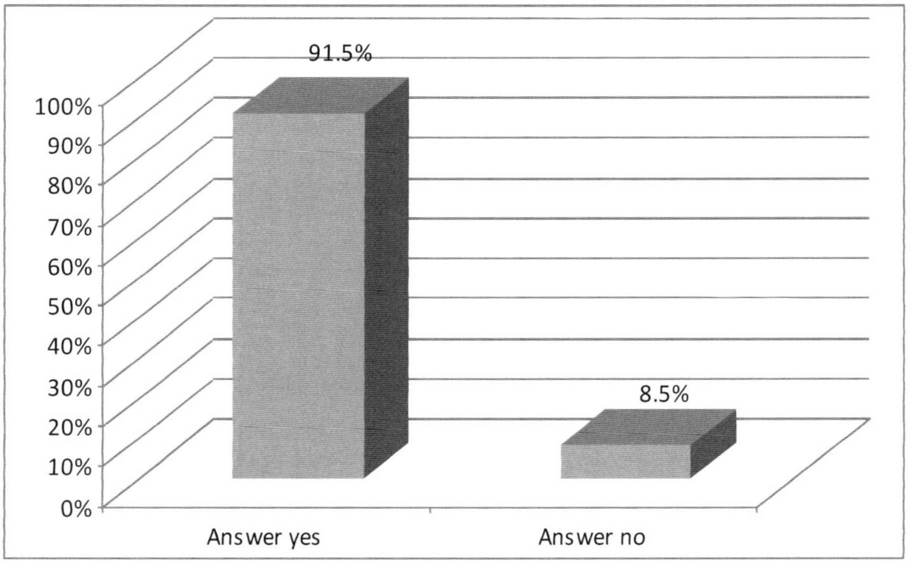

Figure 10. Declared answers to question No. 10.

Job satisfaction is an important element that affects work performance as well as relationships within a team. The respondents in 92.6% have declared satisfaction with their work, some rescue work treats it a passion in which they save lives that are irreplaceable (figure 11).

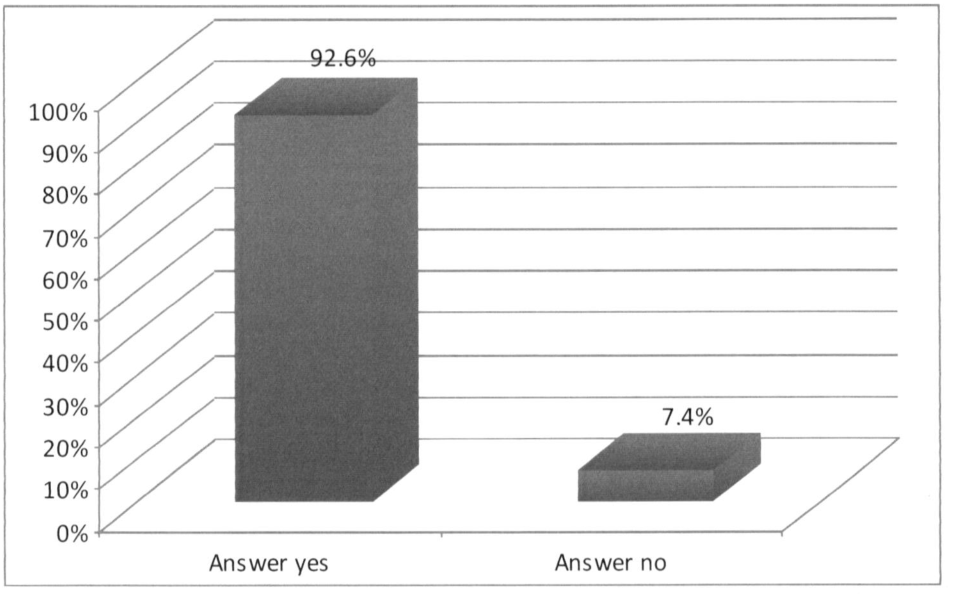

Figure 11. Declared answers to question No. 11.

Rescue workers are very well prepared not only mentally but also physically. The question 'Do you feel that you are doing your work on the edge of endurance?'. Have confirmed perfect preparation because only 3.2% of respondents have the feeling that they are working at the limit of endurance (figure 12).

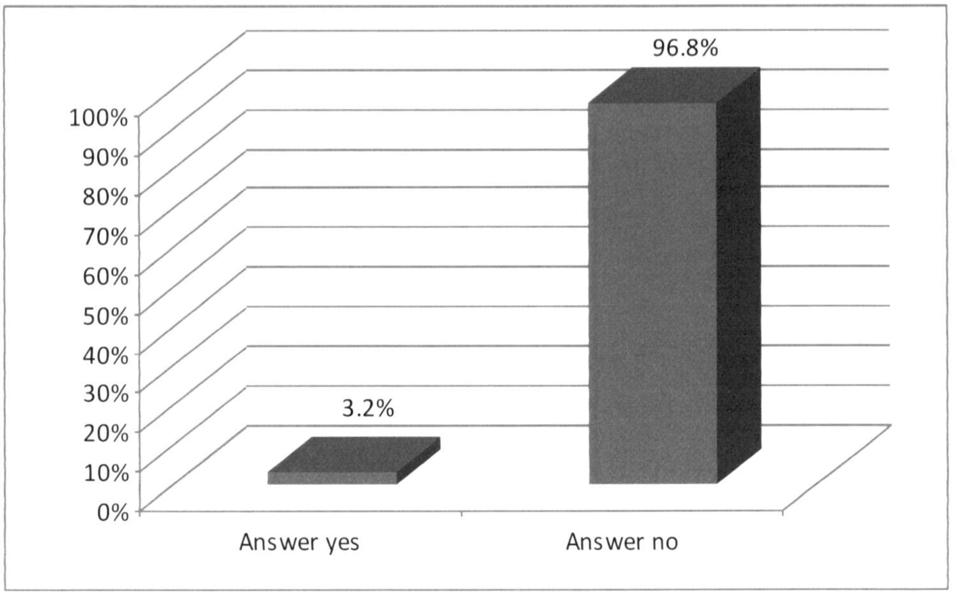

Figure 12. Declared answers to question No. 12.

Based on the research conducted on the importance of work in the life of the rescuer, the authors proposed categories of validity of motives present in the life of the rescuer:

- Category I - motivation to achieve the goal of the fastest removal of victims.
- Category II - acceptance by your teammates.

- Category III - appropriate work comfort taking into account its load.
- Category IV - motivation to achieve the goal of the fastest completion of the action.
- Category V - support for colleagues and superiors.
- Category VI - sense of security by performing by the life-saver itself safety-related activities.
- Category VII - satisfaction with your work.
- Category VIII - dominance of favourite activities by the rescuer.
- Category IX - creating a safe workplace.
- Category X - satisfaction with monthly salary.
- Category XI - limitation of activities not liked by rescuer.
- Category XII - the need to improve their qualifications for promotion.

Based on the suggested categories of validity of the motives present in the rescuer's work situation, the scheme "highlighted motives of the importance of work in the life of the rescuer" was proposed, which will be discussed during the pilot studies during the rescue training.

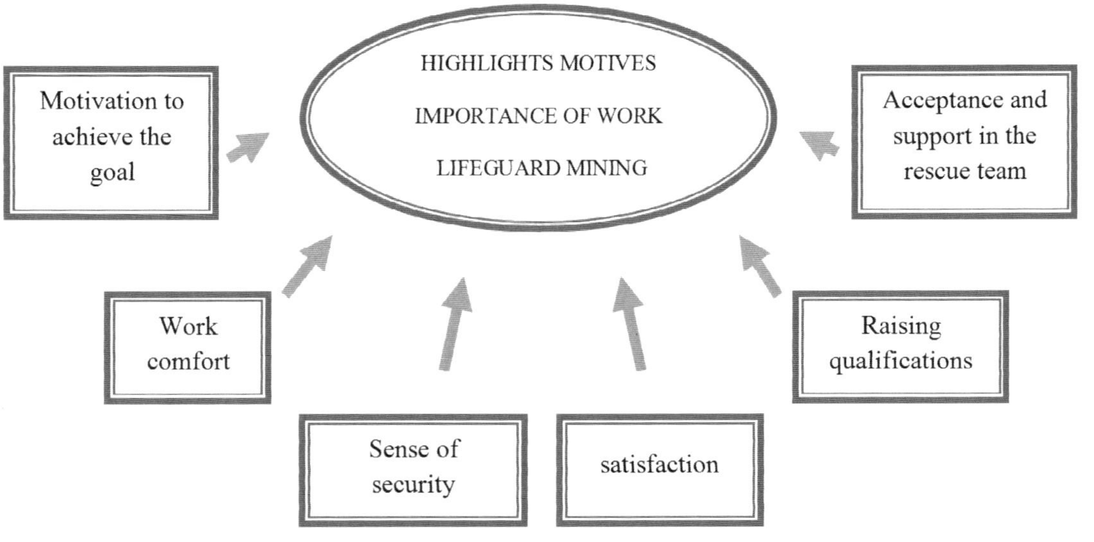

Figure 13. The motives of the importance of work in the life of the rescuer.

3. Summary

- Mining salvage is voluntary, and this is precisely the choice of the miner who has worked for a minimum of a year in a mining plant and additionally has to meet several criteria in accordance with the regulation. Therefore, it can be stated that the survey was carried out by employees with professional experience. They declared that 95.7% of the action motivates them to achieve the goal of the quickest completion of the action and 100% of the quickest way out.
- Each work, and its exact execution, involves paying for it. Rescue workers in 75.5% are happy with the salary, which indicates a positive attitude to work because it increases the comfort of life. In addition, 66% of lifeguards admitted to the need to improve their qualifications.
- The rescuers declared 99% of the acceptance of their colleagues, and 96% could count on support from their colleagues and superiors. Such results are very positive, which, through appropriate cooperation and support will certainly affect the safety at the workplace.
- Satisfaction with their work was expressed by 92.6% of rescuers surveyed, which has an impact on the importance of work in the mine rescue service with the specificity of teamwork.

- Analyzing the survey results, 12 categories of importance of motives present in the life of the rescuer were proposed. In addition, the scheme 'highlighted themes of the importance of work in the life of the rescuer' was proposed, which will be discussed during the pilot study during rescue training.

References

[1] Warr P et al 2001 Encyklopedia Blackwella *Psychologia Społeczna Jacek Santorski & CO* pp 411-2
[2] Ratajczak Z 2008 *Psychologia pracy i organizacji* (Warszawa: Wydawnictwo Naukowe PWN) pp 42-3
[3] Ratajczak Z 1998 Motywy pracy *Encyklopedia Psychologii Fundacja Innowacja* pp 442-6

IOP Conf. Series: Materials Science and Engineering **268** (2017) 012015 doi:10.1088/1757-899X/268/1/012015

Study of the 'lifeline' as the measure allowing for safe self-rescue of miners in conditions of lack of visibility caused by underground fire

Henryk Badura, Aneta Grodzicka and Dariusz Musioł

Faculty of Mining and Geology, Silesian University of Technology, Akademicka 2, 44-100 Gliwice, Poland

henryk.badura@polsl.pl

Abstract. The article presents statistical data regarding the evacuation of miners affected by underground fire hazards. The data indicates that the hazard remains considerable. Due to the increasing lengths of escape routes, measures should be introduced in the longwall regions, which shall improve the safe evacuation of miners, especially in conditions of highly limited visibility or the lack thereof. Within the research project No. 12, entitled: 'The development of orientation systems and systems for signalling the direction of crew withdrawal for escape routes in longwall gates', which is a part of the Strategic Research Project entitled 'The improvement of work safety in mines' financed by the National Centre for Research and Development, a 'lifeline' was designed – that is, a measure that is very effective in the evacuation of staff. Subsequently, tests of time of passage were conducted in the 'Krupiński' coal mine in conditions of lack of visibility. The tests have confirmed the suitability of the 'lifeline' as the measure used for orientation towards the correct direction of evacuation, which increases the pace and the confidence while travelling through the escape route. The mean speed of passage through the heading with an upwards inclination of 11° was 22 m/min, while in case of a nearly horizontal longwall gate, it was 39 m/min.

1. Introduction

Fire hazard is one of the main hazards occurring in hard coal mines. Currently, a large number of outbreaks of spontaneous fires constitutes a problem. Exogenous fires, usually more dangerous than the spontaneous fires, are currently rather rare. This is not only related to the considerable limitation of flammable substances held in underground workings, the prohibition against using open fire, but also due to the appropriate design of the preparatory headings system in the region of the exploitation field [1]. The outbreak of a fire it is most often manifested by the increase of concentration of fire-related gases in the mine's atmosphere [2, 3], especially carbon monoxide.

In the years between 1990 and 2014, 8 exogenous and 33 spontaneous fires occurred in hard coal mines [4].

In case of occurrence of a fire hazard in the longwall region, the staff should be withdrawn from the endangered region using escape routes to a location with streamlined air that is not affected by fire-related gases.

Escape routes are specified in the Mine Rescue Plan, in line with the Ordinance of the Minister of Economy dated June 28th, 2002 [5] and the Ordinance of the Minister of Economy dated June 12th, 2002, regarding the mine rescue operations (Journal of Laws No. 94, pos. 838) [6].

In line with the ordinance [5], 'the highest allowable length of routes with independent air current shall be determined given consideration to the time of operation of the applied respiratory protection measures'. This general wording, which does not directly state the conditions that must be fulfilled by the escape route, only indicates the fact that while designing the exploitation in a given location of the deposit, one should note the necessity of ensuring the safe withdrawal of the crew in the event of the occurrence of a fire.

The efficiency of the escape route is dependent on many technical and human factors [7, 8, 9]. The technical factors include:

- the geometrical parameters of the escape route heading,
- obstacles present in the headings (devices, belt conveyors, pipelines and equipment),
- the direction of withdrawal – the upwards or downwards inclination angle,
- the distribution of communication devices,
- the marking of the escape route,
- the escape respirators.

The human factors include [7, 8, 10]:

- the ability to use the escape respirator,
- the ability to proceed through the escape route,
- the resistance to stress related to conditions occurring in the heading due to the fire,
- the ability to make decisions during the staff self-rescue operation.

Within the first stage of the research project No. 12, accident reports and hazard occurrence reports were compiled and analyzed while giving special consideration to the division into exogenous and spontaneous fires. The analysis encompassed the period from 1990 to 2013.

In the period in concern, per 61 cases of evacuation related to fires, as many as 42 were related to spontaneous fires, while only 19 were caused by exogenous fires (figure 1). The highest number of spontaneous fires (17) occurred in 2009 and 2010, while the highest number of exogenous fires (4) occurred in 2008. In 1990 and in 2006 no fire hazard causing the evacuation of miners was noted.

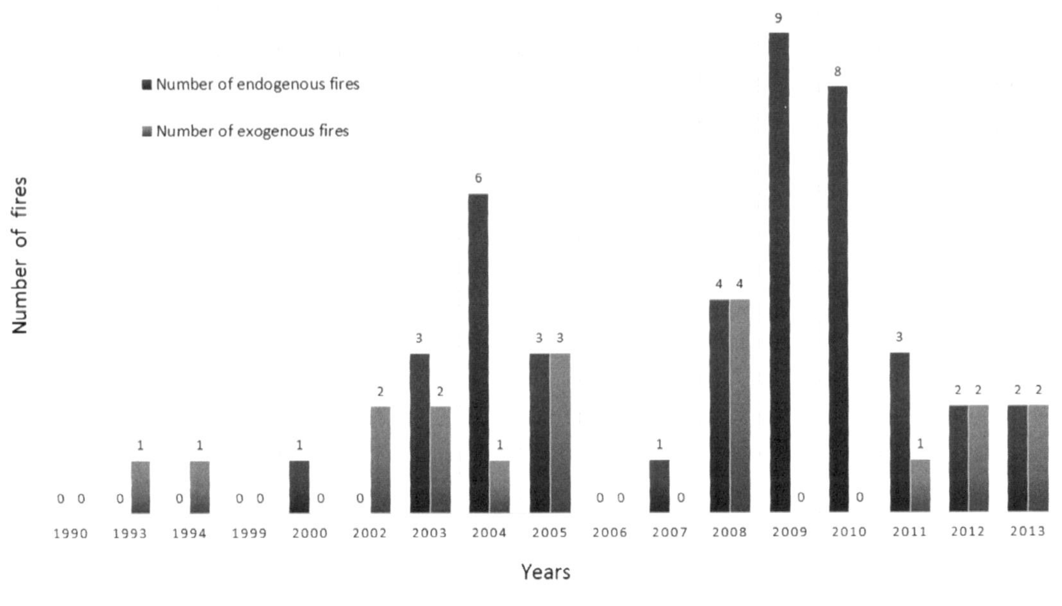

Figure 1. Number of crew evacuations due to fire hazard [11].

The conducted analysis of crew evacuations caused by fire hazards (figure 2) in the years 1990-2013, exhibited that a total of 2574 people were evacuated from the affected headings: 1312 people were evacuated due to spontaneous fires and 1262 people were evacuated due to exogenous fires. The highest number of people evacuated during a single operation was reached in 2005 in the 'Bielszowice' hard coal mine. The exogenous fire resulted from a short-circuit in a 6 kV cable duct at the surface and connected with the downcast shaft. Due to that, 488 people were evacuated from the underground headings.

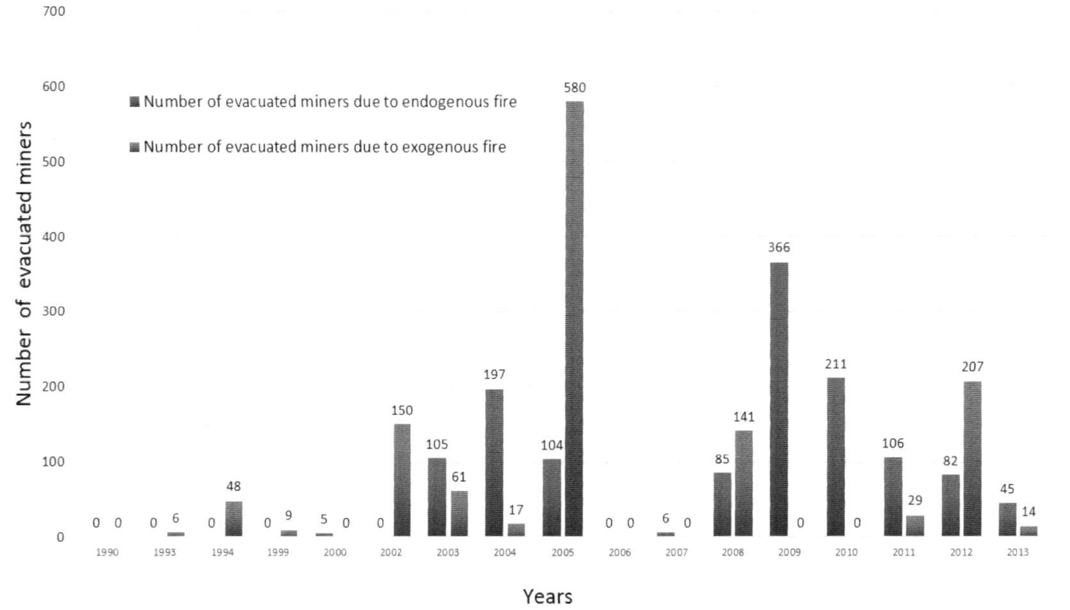

Figure 2. Number of people evacuated due to a fire hazard [11].

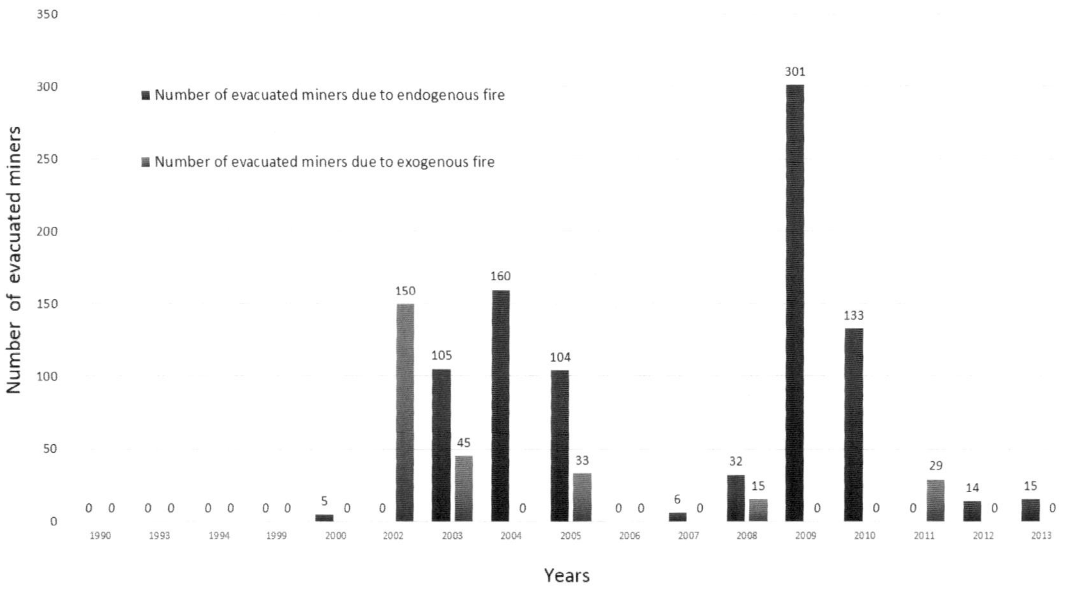

Figure 3. Number of people evacuated from longwall regions due to fire hazard [11].

Among all 61 occurrences related to fire hazards in hard coal mines, as many as 34 [11] concerned the longwall areas, while only 6 of them were exogenous fires. The fires which occurred in the longwall regions constitute 56% of all fires in hard coal mines. The total number of people evacuated from the region of longwalls was 1147 (figure 3). Out of this number, 272 people were evacuated due to exogenous fires and 875 people were evacuated due to spontaneous fires.

According to accident protocols [12], during the development of a fire, both exogenous and spontaneous, considerable smoke often occurred in the headings which constituted the escape route from the affected region. The smoke led to a significant decrease in visibility. Due to the above, there is a necessity to equip the escape routes in measures allowing the withdrawal of the crew in the correct direction. The binding mining law contains a provision stating that the escape routes should be 'appropriately marked' [5], while it does not state which measures should be used. Currently, mines apply various methods for marking the escape routes and the direction of crew evacuation. Metal signs with the name of the heading and the direction of the evacuation through the escape route are most often applied. Such marking only serves its purpose in case of good visibility, while during the occurrence of dense smoke, it would not be visible.

Due to the above, a universal system for the orientation and direction of crew in case of evacuation by means of escape routes should be introduced. Such system should allow for the choice of the correct direction of evacuation, even in conditions of complete lack of visibility. The conducted analysis of Polish mining literature has exhibited that no non-electronic system serving the orientation and signalling the withdrawal direction in high smoke density conditions that completely or highly limits the visibility, is used in the Polish coal mines.

The complete lack of visibility or a high limitation thereof, causes the evacuated crew to only decide on their direction of evacuation based on sound signals or their sense of touch.

Figure 4 exhibits the number of accidents which occurred during the withdrawal of crews from the longwall regions. A total of 36 fatal accidents occurred. It is worth noting that all the fatal accidents were caused by the combustion or explosion of methane.

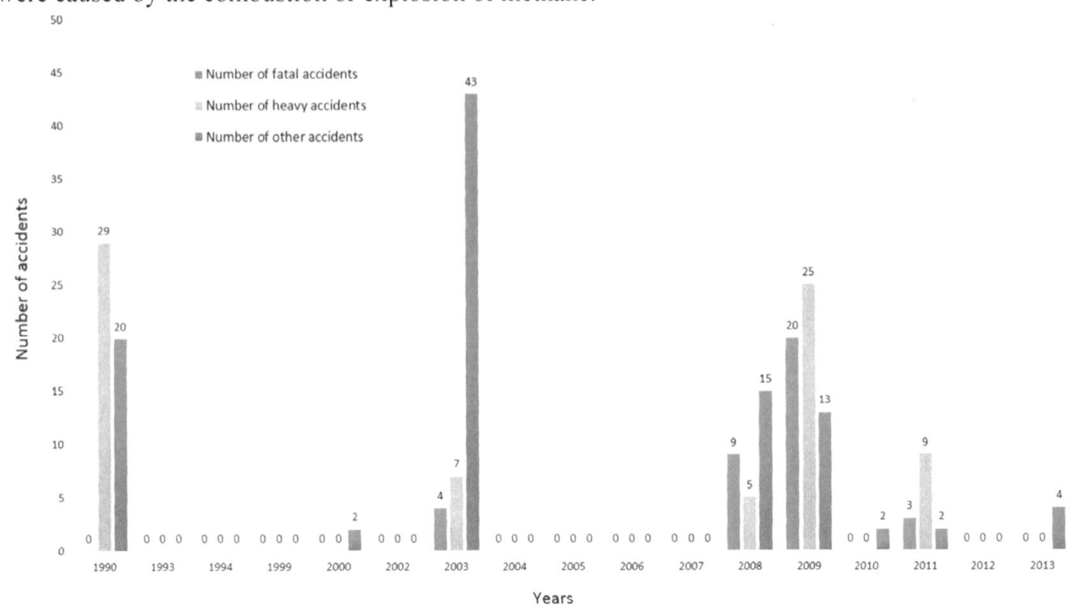

Figure 4. Number of accidents in longwall regions related to the hazardous incidents.

Two accidents occurred in 2003:

- 'Brzeszcze' hard coal mine – methane was ignited due to an exogenous fire, which resulted in 1 fatal accident,

– 'Sośnica' hard coal mine – methane was ignited due to a spontaneous fire, resulting in the death of 3 employees.

In 2008, also 2 accidents occurred – the ignition and explosion of methane in 'Borynia' coal mine resulted in 6 fatalities, while the spontaneous fire in 'Mysłowice-Wesoła' coal mine caused the ignition of methane and the explosion of coal dust, which resulted in two fatal accidents (1 employee died during the explosion, the second one died in hospital due to injuries).

In 2009, the ignition and explosion of methane in the 'Wujek' hard coal mine, Śląsk section, caused the death of 20 miners.

In 2011, the combustion of methane in the 'Krupiński' coal mine resulted in 3 fatalities.

2. 'Lifeline' as an element supporting the evacuation of staff from the affected region

In the research project No. 12 of the Strategic Research Project named 'Improvement of work safety in mines' financed by the National Centre for Research and Development, the so-called 'lifeline' was proposed as a measure used for the orientation in terms of the correct direction of staff evacuation. The notion is from literature as a measure supporting staff evacuation in mines and is related with Anglo-Saxon mining tradition. In the available literature, however, the structure of the 'lifeline' has not been described in detail.

The structure of the 'lifeline' should be simple and allow for easy memorization of markings. It was proposed that only three types of markers should be used: a cone, a cylinder and a sphere (figure 5). All the markers exhibit a similar diameter of 80 mm. The length of the cylinder and the cone is 120 mm.

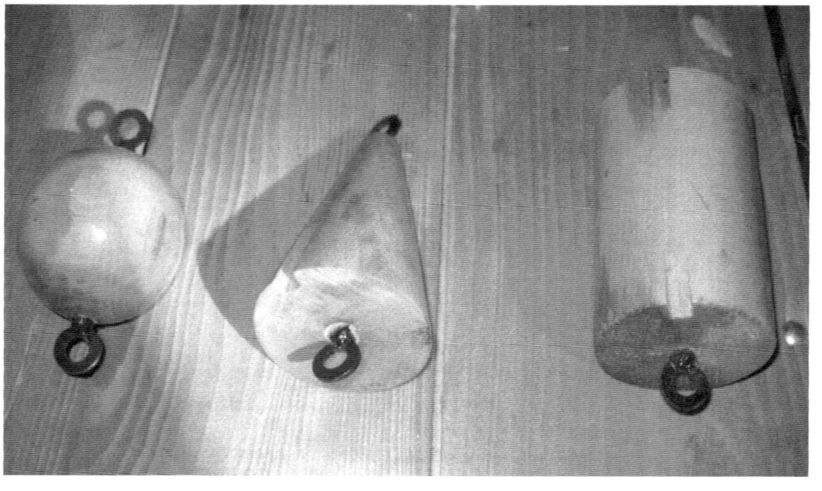

Figure 5. Markers used in the 'lifeline' [11].

The cone indicates the direction of evacuation along the heading. This relates to the graphical representation of direction – an arrow. The cone plays the role of the arrow-head showing the evacuation direction – one should proceed from the base towards the direction shown by the tip.

The cylinder denotes the passage through a crossing in line with the direction of the longitudinal axis of the cylinder. The sphere denotes attention required in relation to further marking or obstacles on the escape route (such as a platform over water).

The sphere also precedes the cylinder which denotes the crossing.

The requirement to move to the other side of the heading is denoted by two cones connected by their tips. The cones are preceded by a sphere, which requires attention to further marking.

A larger amount of spheres denotes a hindrance in the passage such as a low suspended railway, temporary flooding etc. For a person using the escape route it denotes that they should pay attention to all possible hindrances assembled at a small length of the heading.

The 'lifeline' should be suspended at a height of approximately 1.1 m from the floor level, that is, at a height similar to rails in buildings. The evacuating people should move their hands along the 'lifeline' while reading the markers on the way.

It was proposed that the distance between the markers indicating the direction of evacuation at straight sections of the headings should be approximately 5 m. While approaching the markers indicating a change in direction, the density is increased to 2 m and subsequently to 1 m.

The combinations of the markers exhibited below allow for conveying the information concerning navigating through the escape route.

A passage to the other side of the heading is denoted by two truncated cones directed towards each other with the narrower end and preceded by a sphere indicating the necessity to increase attention. The arrow in the figure (figure 6.) denotes the direction of the crew evacuation.

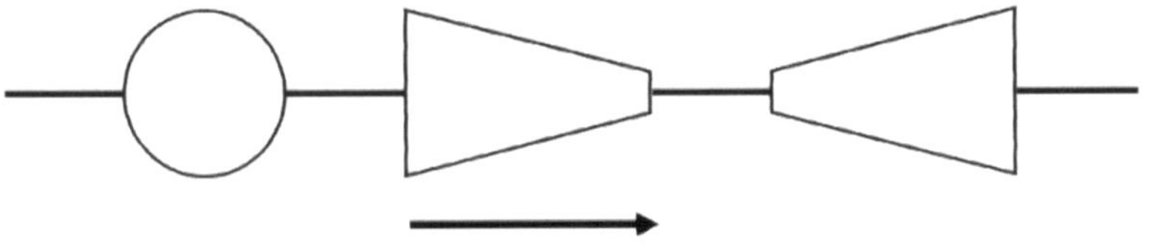

Figure 6. Passage to the other side of the heading [11].

Such a combination of markers should be read as the following order:

Stop. Turn back to the 'lifeline'. Move to the other side of the heading.

Shall the miner not notice the above combination of markers, a meter further a pair of markers is placed which informs them of the wrong direction of evacuation. That group of markers is presented in figure 7.

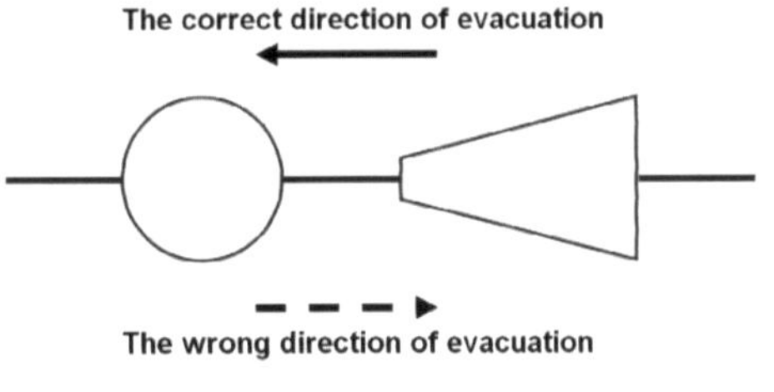

Figure 7. Change in the direction of evacuation [11].

A complete combination of markers denoting the passage to the other side of the heading including the information regarding the wrong direction of evacuation has been presented in figure 8.

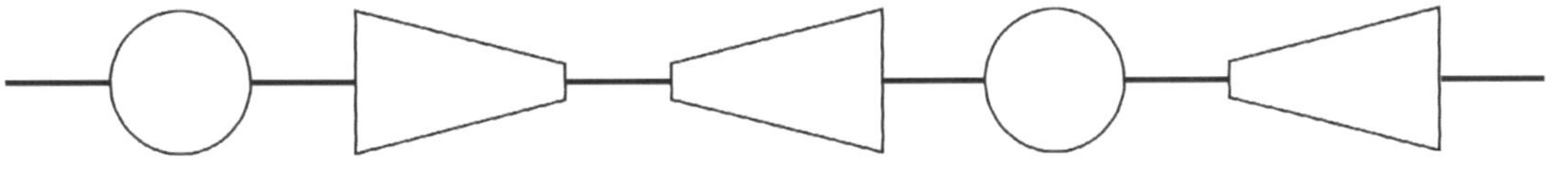

Figure 8. Marking denoting the passage to the other side of the heading [11].

During the evacuation along the escape route, it might be necessary to go straight at a crossing. For that purpose, a combination of the sphere and cylinder markers should be suspended on the 'lifeline'. Such a combination was presented in figure 9. The arrow denotes the direction of passage.

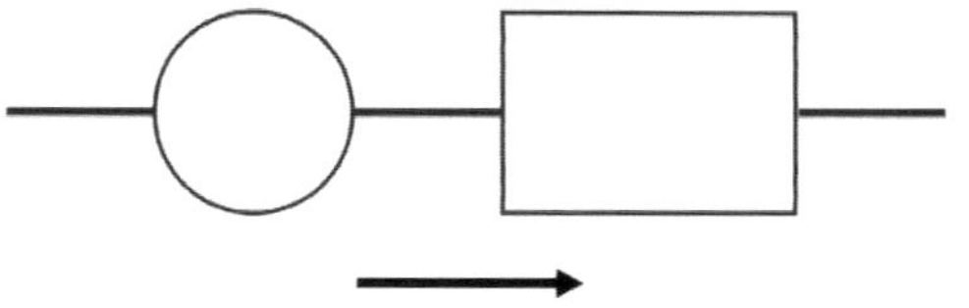

Figure 9. Marking denoting the necessity to go straight at a crossing [11].

An example of markers at a 'lifeline' has been presented in figure 10.

The drawing shows a case of evacuation using the escape route on the right side of the heading.

The miners proceed along the direction of the flow of smoke. During the evacuation they find markers denoting the necessity to pass through a crossing on the 'lifeline'. After passing the crossing, they find another 'lifeline', providing the further direction of evacuation. After passing a certain section of the heading, the miners find another combination of markers on the 'lifeline', which informs them on the necessity to pass to the other side of the heading.

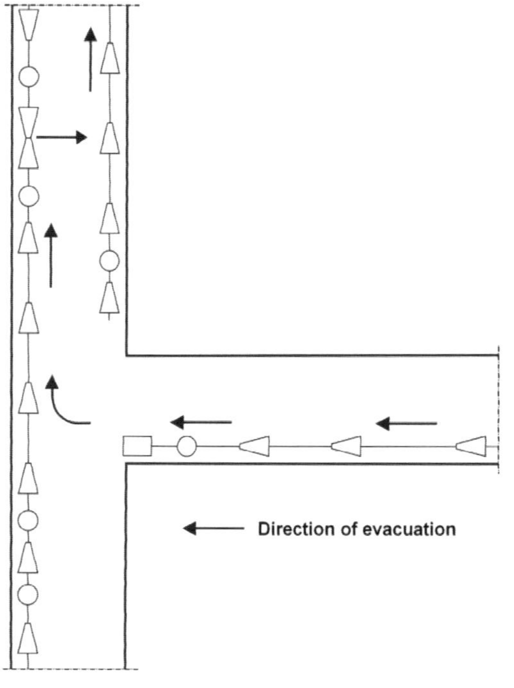

Figure 10. An example of application of the 'lifeline' in an escape route [11].

An example of a 'lifeline' installed in a heading has been shown in figure 11. It illustrates a combination of markers requiring an increased attention (sphere), and subsequently the requirement to move to the other side of the heading (two cones with tips directed towards each other). Subsequent markers inform the miner on the wrong direction of evacuation and the order to move back to the two cones directed towards each other.

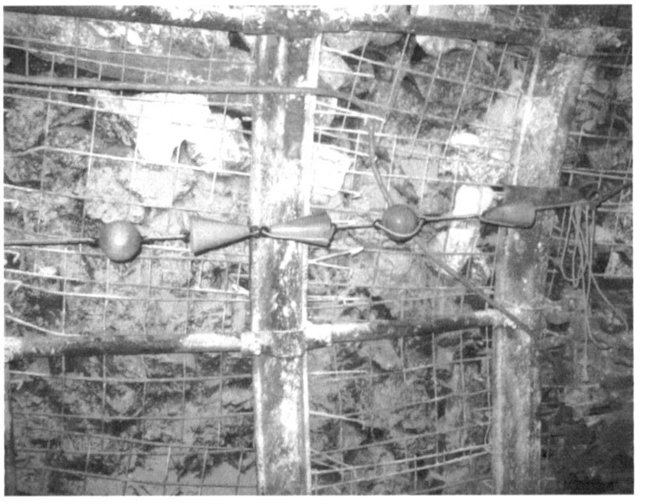

Figure 11. A combination of markers ordering the passage to the other side of the heading [11].

3. Verification tests using the 'lifeline'

Three stages of verification tests were conducted in underground headings of the 'Krupiński' coal mine. The tests were conducted in conditions of simulated loss of visibility. The complete loss of visibility was simulated using special glasses.

The first stage of the tests was conducted in a technological drift and in a collective ventilation roadway. The total length of the evacuation route was 100 m. Figure 12 presents the diagram of the evacuation using the escape route with the 'lifeline' suspended in the headings of the auxiliary ventilation road.

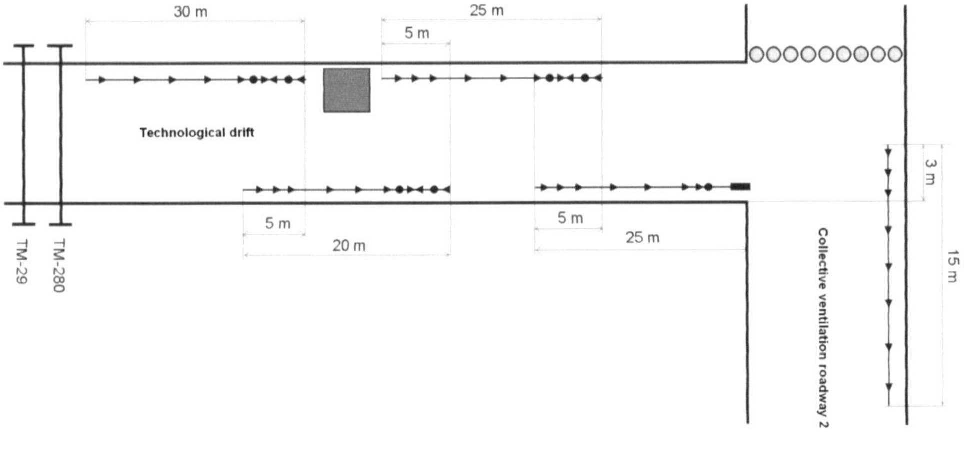

Figure 12. The placement of the 'lifeline' in the ventilation road in the first stage of measuring the speed of passage [11].

The first stage of the tests encompassed 22 miners employed by the mine. The length of the route was 100 m, including a 90 m section of drift with an inclination angle of 10° and a section of 10 m of a horizontal gallery. Table 1 presents the passage times and the speeds of passages of the rescue route using the 'lifeline'. The mean speed was 22.0 m/min. The minimal speed was 13.4 m/min, while the maximal speed was 34.7 m/min.

Table 1. Passage times and speeds of the escape route using the 'lifeline' in stage I of the tests [11].

Worker No.	Time of passage (min:sec)	Speed (m/min)
Worker No.1	4:01	24.9
Worker No.2	4:06	24.4
Worker No.3	3:38	27.5
Worker No.4	3:40	27.3
Worker No.5	3:25	29.3
Worker No.6	3:31	28.4
Worker No.7	6:31	15.3
Worker No.8	4:23	22.8
Worker No.9	7:27	13.4
Worker No.10	5:26	18.4
Worker No.11	6:33	15.3
Worker No.12	7:08	14.0
Worker No.13	5:53	17.0
Worker No.14	4:30	22.2
Worker No.15	4:41	21.4
Worker No.16	5:16	19.0
Worker No.17	4:59	20.1
Worker No.18	5:41	17.6
Worker No.19	6:32	15.3
Worker No.20	4:20	23.1
Worker No.21	2:58	33.7
Worker No.22	2:53	34.7
Mean time of passage	4:53	
Minimal time of passage	2:53	
Maximal time of passage	7:27	
	Mean speed	22.0
	Minimal speed	13.4
	Maximal speed	34.7
	Standard deviation	6.2

In the second stage of the experiment, the 'lifeline' was suspended partially in a technological drift and partially in the collective ventilation road No. 2. The placement of the lifeline is presented in figure 13. The length of the line was 135 m and the route was 110 m. The tested group was comprised of 10 people including 5 mine rescue workers and 5 employees of the Institute of Mining of the Silesian University of Technology in Gliwice. The trial of passing the escape route was conducted twice. The results of the measurements have been presented in table 2.

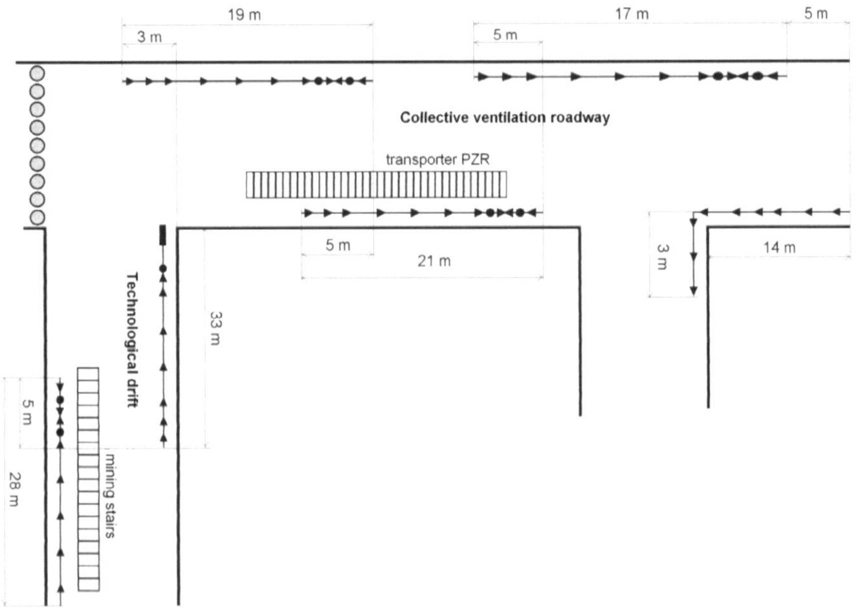

Figure 13. The placement of the 'lifeline' in the ventilation road in the second stage of measuring the speed of passage [11].

Table 2. Passage times and speeds of the escape route using the 'lifeline' in stage II of the tests [11].

No.	Worker No.	1st time of passage (min:sec)	1st speed of passage (m/min)	2nd time of passage (min:sec)	2nd speed of passage (m/min)
1.	Worker No.1	5:29	18.2	3:45	26.7
2.	Worker No.2	2:48	35.7	2:13	43.8
3.	Worker No.3	4:45	21.1	2:53	34.7
4.	Worker No.4	4:16	23.4	2:55	34.3
5.	Worker No.5	7:53	12.7	3:45	26.7
6.	Mine rescue worker No.1	2:58	33.7	2:56	34.1
7.	Mine rescue worker No.2	3:04	32.6	2:02	49.2
8.	Mine rescue worker No.3	2:53	34.7	2:42	37.0
9.	Mine rescue worker No.4	2:38	38.0	2:12	45.5
10.	Mine rescue worker No.5	2:17	43.8	2:11	45.8
	Mean time of passage	3:54		2:45	
	Minimal time of passage	2:17		2:02	
	Maximal time of passage	7:53		3:45	
	Mean speed		29.4		37.8
	Minimal speed		12.7		26.7
	Maximal speed		43.8		49.2
	Standard deviation		9.9		8.0

All people participated in the test for the first time and did not know the route indicated by the 'lifeline'.

During the first passage, the employees of the Silesian University of Technology reached speeds that were significantly lower than that of the 'Krupiński' mine rescue team members. Their passage speeds were in the range from 12.7 m/min up to 35.7 m/min, while the rescuers were moving at speeds from 32.6 m/min up to 43.8 m/min. The mean speed of passage calculated for all the participants of the experiment was 29.4 m/min.

The speeds achieved by each of the participants in the second passage were higher than in the case of the first passage. A larger change was noted in the group of the Silesian University of Technology's employees. In case of this group the speeds were in the range from 26.7 m/min up to 34.7 m/min. The passage speeds of the rescuers was in the range from 34.1 m/min up to 49.2 m/min. The mean value of speed calculated based on individual speeds of each of the participants of the experiment was 37.8 m/min.

The considerable increase in the speed of passage in the second stage indicated a significant impact of the knowledge of the escape route on the speed. This means that bringing the crew out through escape routes significantly increases the fire safety in a mine.

The third stage of the passage speed measurement in conditions of lack of visibility was conducted in the N-6 top gate of the N-8 longwall in the 330/2 seam. The length of the 'lifeline' was 210 m. The line was comprised of three equal sections. The first and the last one was on the right side of the heading while the middle one led through the left side of the heading. Figure 14 presents the cross-section of the N-6 gate of the N-8 longwall.

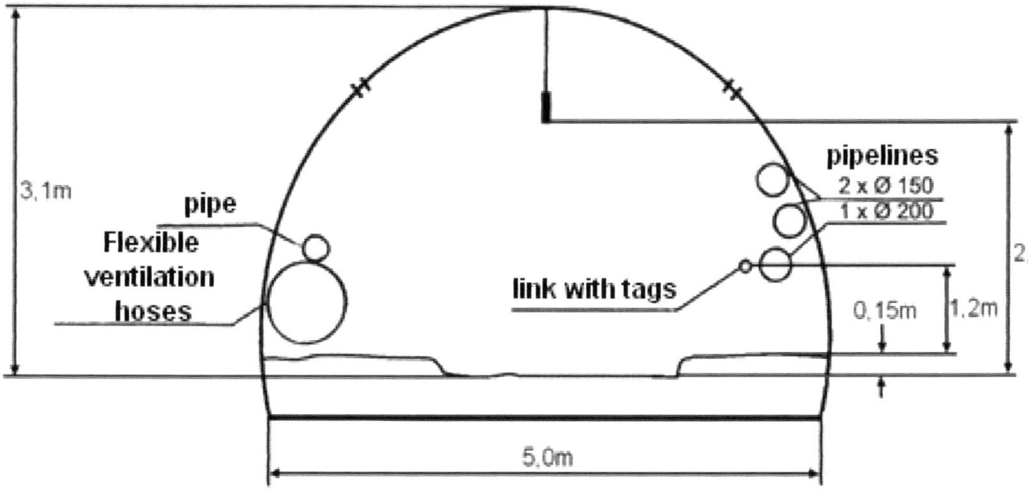

Figure 14. Cross-section of the N-6 gate of the N-8 longwall in the 330/2 seam at the 'Krupiński' coal mine [11].

19 miners, including 6 rescue team members, 12 members of the crew working at the longwall and one employee of the Silesian University of Technology were subjected to the test. All participants of the experiment used KA-60 type escape respirators.

The mean speed of passage was 39.0 m/min. The minimal speed of passage was 21.8 m/min while the maximal speed was 53.8 m/min. The coefficient of variation (a ratio of the standard deviation and the mean speed expressed as a percentage) was 18.4%, which means that the group of participants was relatively uniform.

The experiment received large interest and approval of the participating miners.

Table 3. Passage times and speeds of the escape route using the 'lifeline' in stage III of the tests [11].

Worker	Time of passage (min:sec)	Speed of passage (m/min)
Worker 1	9:11	21.8
Worker 2	5:31	36.3
Worker 3	5:38	35.5
Worker 4	5:06	39.2
Worker 5	4:34	43.8
Worker 6	5:16	38.0
Worker 7	3:43	53.8
Worker 8	4:55	40.7
Worker 9	5:47	34.6
Worker 10	5:05	39.3
Worker 11	5:05	39.3
Worker 12	6:00	33.3
Worker 13	4:39	43.0
Worker 14	4:33	44.0
Worker 15	3:58	50.4
Worker 16	5:22	37.3
Worker 17	4:44	42.3
Worker 18	4:54	40.8
Worker 19	7:11	27.8
Mean time of passage	5:19	
Minimal time of passage	3:43	
Maximal time of passage	9:11	
	Mean speed	39.0
	Minimal speed	21.8
	Maximal speed	53.8
	Standard deviation	7.2

4. Summary

Underground fires in coal mines occur despite the considerable progress in fire hazard prevention. Such fires constitute a high risk for the crew and may cause extensive material losses.

Currently, fires caused by methane ignition are considered to be the most dangerous. Shall such a fire occur in the goafs in the vicinity of a longwall, the fire develops rapidly, a considerable increase in temperature of the flowing air is observed and other substances are combusted (such as coal, wood, bonding agents, conveyor belts etc.), which, in turn, results in production of large amounts of fumes. A methane fire in goafs may not be extinguished and the only method to save lifes is to evacuate the affected headings as fast as possible. In dense smoke, the visibility is largely limited and the conditions in the heading force the miners to withdraw from the affected zone as soon as possible.

At a highly limited visibility, or the lack thereof, the evacuating miners may lose orientation regarding the correct direction of evacuation. Especially the horizontal or just slightly inclined headings as well as crossings of headings are dangerous in that respect. In conditions of dense smoke the miners may only use their senses of touch and hearing.

The conducted research exhibited that the 'lifeline' may be a very helpful measure used in the evacuation from the hazard zone in conditions of a complete loss of visibility.

Based on the values of passage speed through the headings obtained in experiments, it may be stated that the 'lifeline' has considerably increased the speed of passage in conditions of complete lack of visibility.

The 'lifeline' should be suspended in such a manner that it would lead the evacuating miners to communication devices and escape respirator replacement chambers.

The arrangement of the 'lifeline' in crossing areas should require the verification whether air without smoke reaches the crossing. The 'lifeline' has received a very positive reception by the miners participating in the experiments conducted in the 'Krupiński' coal mine.

The markers used in the 'lifeline' should be equipped with reflective strips facilitating the orientation in terms of the evacuation direction and allowing to find where the lifeline is suspended.

Evacuation of the staff in a direction opposite to the airflow direction is usually conducted in an air stream not containing smoke. The evacuation of the crew in a direction opposite to the flow of smoke may only be applied when it reduces the length of the escape route and it excludes the possibility of reaching the centre of the fire.

References

[1] Sułkowski J and Musioł D 2008 Effect of bed splitting on fighting aerologic hazards in exploitation sections of hard coal mines *Archives of Mining Sciences* **53/4**

[2] Dziurzyński W, Krach A and Pałka T 2010 Computer simulation of the impact of fed-in mineral substances on air and methane flow in longwall goaf *Archives of Mining Sciences* **55/3**

[3] Dziurzyński W, Krach A and Pałka T 2014 Computer simulation of the propagation of heat in abandoned workings insulated with slurries and mineral substances *Archives of Mining Sciences* **59/1**

[4] Ocena stanu bezpieczeństwa pracy, ratownictwa górniczego oraz bezpieczeństwa powszechnego w związku z działalnością górniczo-geologiczną w 2013 r. Opracowanie Wyższego Urzędu Górniczego w Katowicach. http://www.wug.gov.pl/download/5477.pdf

[5] Rozporządzenie Ministra Gospodarki z dnia 28 czerwca 2002 r. (z późniejszymi zmianami) w sprawie bezpieczeństwa i higieny pracy, prowadzenia ruchu oraz specjalistycznego zabezpieczenia przeciwpożarowego w podziemnych zakładach górniczych, Dz. U. Nr 139, poz. 1169.

[6] Rozporządzenie Ministra Gospodarki z dnia 12 czerwca 2002 r. w sprawie ratownictwa górniczego (Dz. U. Nr 94 poz.838)

[7] Badura H, Biernacki K, Sułkowski J and Żur K 1996 Czynniki decydujące o prędkości wycofywania załogi zadymionymi wyrobiskami w czasie pożaru w kopalni *Przegląd Górniczy* **6** pp 265–71

[8] Badura H and Sułkowski J 1996 Droga ucieczkowa jako element systemu zabezpieczenia przeciwpożarowego kopalń *Przegląd Górniczy* **7-8** pp 298–311

[9] Sułkowski J et al 1994 Badania oceny dróg ucieczkowych w kopalniach węgla. Praca zbiorowa Instytutu Eksploatacji Złóż

[10] Sułkowski J. et al 1993 Badania prędkości poruszania się górników wyrobiskami podziemnymi symulującymi drogi ucieczkowe w czasie pożaru. Praca zbiorowa Instytutu Eksploatacji Złóż
i Centralnej Stacji Ratownictwa Górniczego w Bytomiu

[11] Szlązak J. et al 2015 Zadanie badawcze nr 12: Opracowanie systemów orientacji i sygnalizowania kierunku wycofania się załogi na drogach ucieczkowych w chodnikach przyścianowych realizowane w ramach Strategicznego Projektu Badawczego Poprawa bezpieczeństwa pracy w kopalniach http://www.ncbr.gov.pl/programy-strategiczne/poprawa-bezpieczenstwa-pracy-w-kopalniach/

[12] Dokumentacja powypadkowa związana z wycofaniem załóg górniczych drogami ucieczkowymi. Archiwum WUG Katowice 1990-2014

IOP Conf. Series: Materials Science and Engineering **268** (2017) 012016 doi:10.1088/1757-899X/268/1/012016

Forecasts of methane concentration at the outlet of the longwall with caving area - case study

Henryk Badura, Piotr Bańka, Dariusz Musioł and Marek Wesołowski

Faculty of Mining and Geology, Silesian University of Technology, Akademicka 2 44-100 Gliwice, Poland

henryk.badura@polsl.pl

Abstract. This paper presents the characteristics of methane hazard and prevention undertaken in the N-6 longwall of seam 330/2 in "Krupiński" coal mine. On the basis of methane concentration measurements conducted with the use of telemetric system, time series of the average and maximum methane concentration at the outlet of the longwall area were generated. It was ascertained that they exhibit a strong autocorrelation. Based on a series of the average methane concentration, a time series of ventilation methane content was created and a total methane content was calculated with the use of methane flow rate measurements in the demethanization system. It was ascertained that dependence between methane concentration and output on the examined day and on the previous day is weak and also that the dependence between methane concentration and air flow rate is very weak. Dependencies between ventilation methane content, total methane content and demethanization efficiency were also investigated. Based on forecasting models [1] developed earlier by H. Badura, forecasts have been made to predict the average and maximum methane concentrations. The measured values f methane concentration show a high level of accordance with forecasted ones.

1. Introduction

The existing mining regulations [2] require performing forecasts of methane content in longwalls prior to the commencement of mining exploitation. The relevance of these forecasts depends on the accuracy of geological reconnaissance of deposit conditions, the degree of recognition of methane volume in the exploited seam and seams lying above and below it as well as methane volume in gangue, especially in sandstones occurring in the roof and floor rocks [3, 4, 5, 6]. The recognition of methane volume based on measurements is usually limited to seams in which the exploitation was or is conducted. Methane volume in seams occurring between the ones in which methane volume was determined is performed by interpolation. Methane volume of seams occurring below the exploited one is determined by extrapolation of methane volume indications in seams occurring higher or interpolated from measurements made in geological test boreholes. It also happens to be assumed by the forecaster basing on his experience. Also methane volume in sandstones is most often set down on the basis of the experience of a person performing forecast.

Methane contained in rock mass occurs in free and sorbed form (adsorbed and absorbed). In coal seams the largest amount of methane is present as sorbed. Free methane usually does not exceed 10% of the total methane content. Sorbed methane is also present in coal-containing mineral layers. The largest amount of free gas occurs in gravels, sandstones - especially coarse grained ones, and in conglomerates, because these rocks exhibit the highest porosity.

IOP Conf. Series: Materials Science and Engineering **268** (2017) 012016 doi:10.1088/1757-899X/268/1/012016

Methods used to forecast the total methane content in the longwall area assume that degassing of rocks that occur in the roof and floor of the exploited part of seam takes place in the so-called desorption zone, the shape and range of which is differently determined by the authors of forecasting methods [7, 4, 8, 5, 9, 10]. In Poland at the present time the most often used theory is the one proposed in papers [6, 7]. It is also assumed that the degree of degassing of layers depends on the distance of the investigated layer from the exploited seam [4, 5, 6, 8]. In Poland, the most commonly used method is the one developed at the Central Mining Institute – 'Barbara' Experimental Mine by B. Kozłowski [5, 6]. Forecasts of total methane content in longwall areas also take into account the predicted output size of the longwall.

In the light of the above forecasted value of methane content in the longwall is determined by many factors, the influence of which cannot be fully predicted. Every methane content forecast is therefore burdened with a certain error, the actual value of which can only be determined by comparison with methane content that was present during the mining exploitation.

Forecasted ventilation methane content is calculated as the product of forecasted total methane content and methane removal efficiency achieved or assumed in the given mine.

Ventilation methane content of the longwall area is calculated as the product of methane concentration and the flow rate of air in the longwall area. Upon forecasting the average methane concentration and knowing or assuming the value of the air flow rate, one can calculate the forecasted ventilation methane content value for the designed longwall.

This paper presents the results of an one-day forecast of the average and maximum methane concentration at the outlet of the N-6 longwall in seam 330/2 of the "B" coal mine and the assessment of forecasts errors.

2. Natural and mining conditions of the N-6 longwall area in seam 330/2

Occurring conditions of part of seam 330/2 exploited by the N-6 longwall were identified by five boreholes of which two were drilled from the surface (BS-30/1981, BS-3/1972) and three from underground excavations (G-23/2005, G-24/2005 and G-17/2007) as well as from preparatory excavations made for the N-6 longwall [11]. Fragments of geological profiles from BS-3/1972 and G-23/2000 boreholes are shown in figure 1. Figure 2 presents geological cross-section of seam parallel to the face of the N-6 longwall.

The analysis of geological profiles shows that geological conditions of seam occurrence were very variable. For example, the BS-3/1972 borehole drilled directly above seam revealed dark gray claystone, cracked, 1.80 m thick, light gray sandstone, cracked, 1.60 m thick and gray claystone, strongly sandy, 9.8 m thick. Over these layers occurs the first layer of coal and coal with claystone with a total thickness of 0.8 m. Due to the total thickness of rock layers, one can conclude that coal layer occurs over seam 330/2 at a distance of 13.20 m.

In the G-23/2000 borehole above seam 330/2 occurs claystone, 0.30 m thick, the layer of sandy claystone with intercalations of sandstone, 6.10 m thick, sandstone, 2.20 m thick and the layer of sandy claystone with the thickness of 5.70 m. Over the mentioned layers, i.e. at a distance of 13.0 m, occurs the coal layer with a thickness of 1.00 m.

It follows from the foregoing that rock layers between seam 330/2 and an unnamed carbon layer differ in mineralogical composition, petrographic composition and thickness. Also the carbon layer has a different thickness and the degree of impurity caused by gangue.

Profiles in figures 1 and 2 show that seam 330/2 is divided into layers by gangue partings. The number of layers and their thicknesses vary in dependence of coordinates of the studied geological profile. In BS-3/1972 borehole (figure 1) the seam has a thickness of 3.50 m, with carbon layers accounting for 54% of this thickness. In G-23/2000 borehole (figure 1) coal has a thickness of 1.10 m and forms one layer with high content of claystones in the roof section.

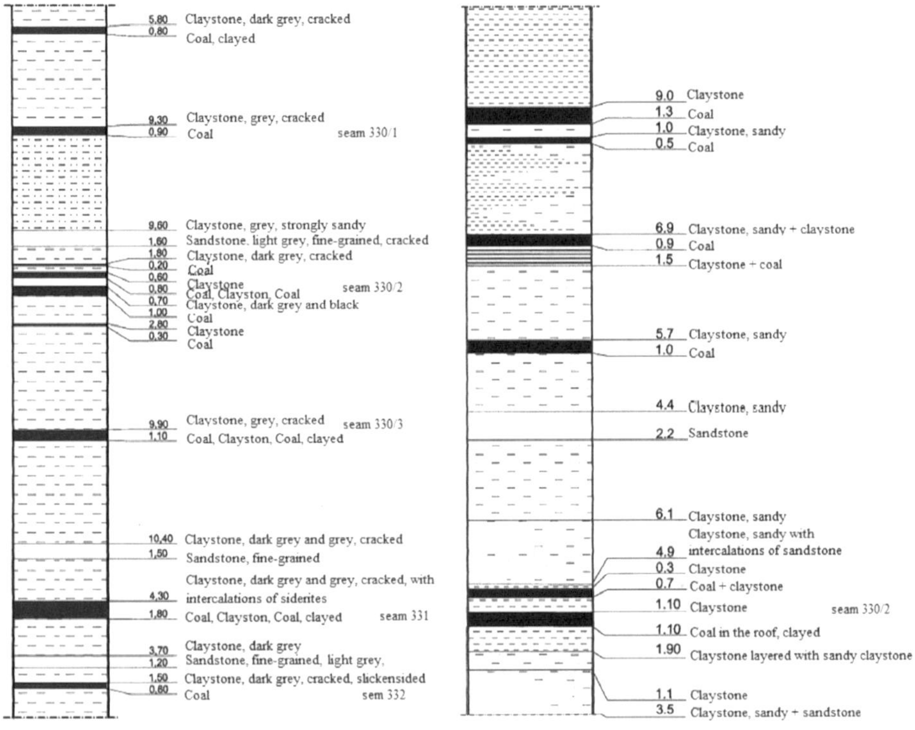

Figure 1. Fragments of geological profiles from BS-31/1972 borehole (left) and G-23/2000 borehole (right).

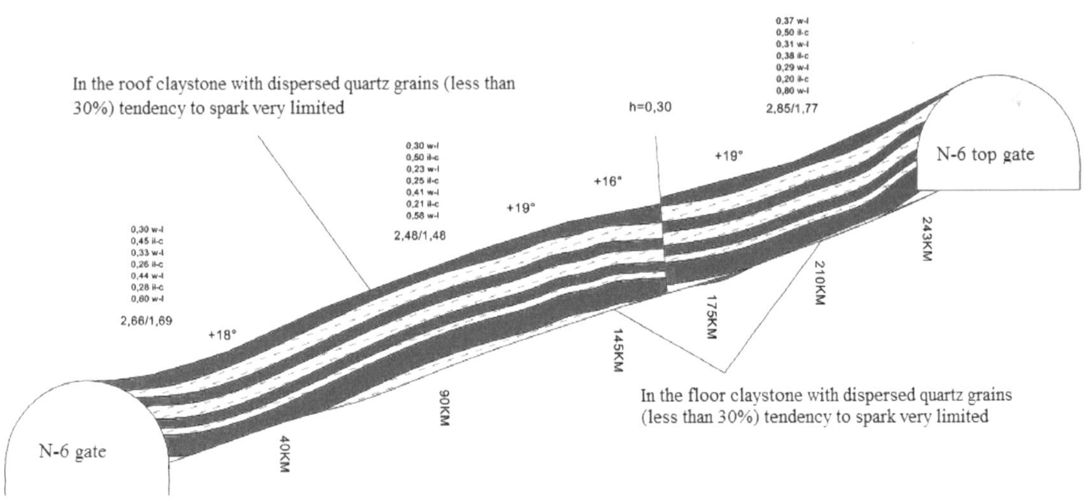

Figure 2. Geological cross-section through seam 330/2 parallel to the longwall face.

In the longwall face profile, in place of the beginning of exploitation with longwall (figure 2) near the N-6 gate, seam has a thickness of 2.66 m and coal layers account for 64% of this thickness (1.69 m). At the point located about 90 m from the N-6 gate, seam thickness is 2.48 m, and the thickness of carbon layers – 1.48 m (60% of seam thickness). At the point located about 210 m from the N-6 gate, seam thickness is 2.85 m and the thickness of carbon layers –1.77 m (62%).

The strength of roof rocks, determined by testing core samples, ranged from 53.5 MPa to 75.1 MPa, the strength of floor rocks ranged from 22.0 MPa to 41.2 MPa. The strength of coal in seam 330/2 ranges from 14.0 MPa to 16.0 MPa.

On the basis of geological profiles, it can be stated that in close proximities to seam 330/2 occur several non-recoverable seams. The deposit also contains mineral layers with considerable content of coal. As mentioned above, such layers are the source of sorbed methane.

The distance of adjacent recoverable seams located over seam 330/2 is variable. The closest earlier exploited seam 329/1, 329/1-2 is located at the distance of 35 m to 75 m.

Forecasted total methane content of the N-6 longwall was 30.14 m^3/min for the planned output of 4000.00 Mg/day. Forecast predicted that approximately 48.6% of total methane content would come from overlying rocks, 26.5% from the exploited seam, and 24.9% from layers occurring below seam 330/2.

The G-6 longwall (figure 3) was characterized by the following geometric parameters:

- longwall length about 225 m,
- panel length about 1100 m,
- height from 2.80 m to 2.96 m.

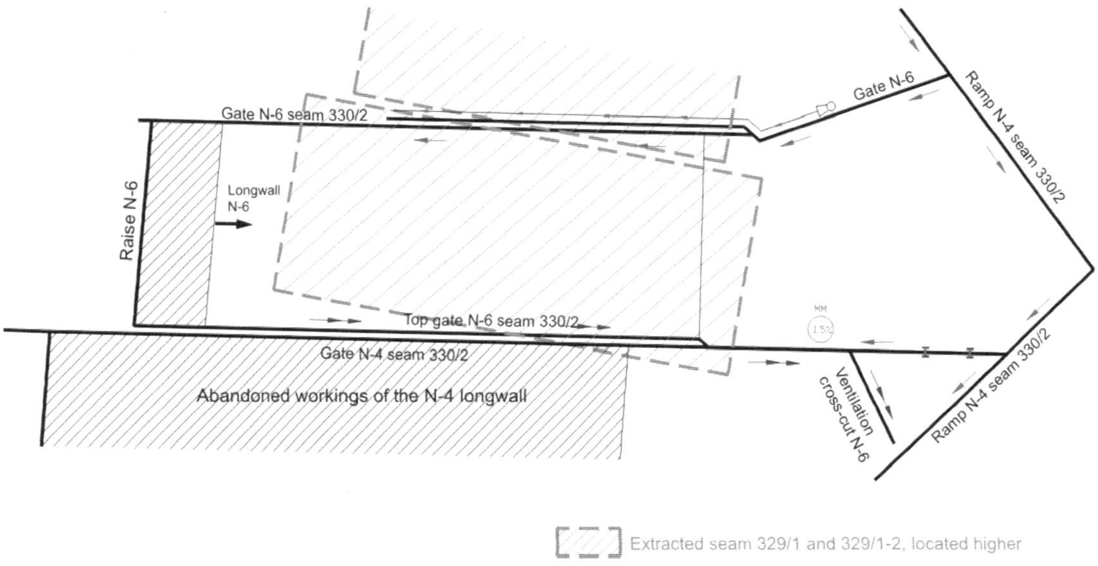

Figure 3. Diagram of workings of the N-6 longwall area in seam 330/2.

In order to prepare the N-6 longwall exploitation area, the following preparatory excavations were made. The N- 6 gate was drilled from ramp N-4 in seam 330/2. From the N-4 gate, a bottom gate for the N-4 longwall, the N-6 top gate was drilled, separated from the N-4 gate by a narrow coal pillar. The N-6 gate and the N-6 top gate were connected by the N-6 raise.

Fresh air from the N-4 ramp in seam 330/2 entered the N-6 longwall through the N-6 gate. After ventilating the longwall, the exhausted air was drained along the N-6 top gate to the N-4 gate. Along the N-4 gate exhausted air flowed into the intersection with the N-4 ventilation cross-cut. Here it merged with fresh air flowing from the N-4 ramp in seam 330/2, sealed by a section of the N-4 gate. From the intersection, the flow of exhaust air was drained by the N-4 ventilation cross-cut to the subsequent excavations towards the ventilation shaft.

In the initial period, at the panel length of about 220 m, the exploitation with the N-6 longwall was carried out under an unexploited part of the deposit. In the remaining part of the exploitation area the longwall entered under the mined out part of seam 329/1, 329/1-2, which is depicted in figure 3 in the form of beige hatched surfaces. As the distance between seams 330/2 and 329/1, 329/1-2 ranged from 35 m to 70 m, the exploitation in seam 329/2, 329/1-2 caused partial demethanization of seam 330/2 and rock layers between these seams. As the exploitation progressed, the length of longwall underneath workings in seam 329/1, 329/1-2 increased.

The average monthly longwall advance was 105 m, i.e. about 5 m per working day.

3. Measurement data

Forecasts of methane concentration were based on telemetric measurements. Methane concentration sensor was located in the exhaust air current in the N-4 gate, in front of the intersection with the N-6 ventilation cross-cut (figure 3). Measurement error of the methane concentration sensor was 0.1% CH_4. Measurements used to create and verify forecasts were conducted in the period from 23.05.2013 to 04.02.2014, i.e. 257 days, including 175 working days. The first day of the observation period was also the first day of the N-6 longwall exploitation.

Measurement data were processed with the use of PROGMET program, developed at the Institute of Mining of Silesian University of Technology [1]. The results of calculations include among others the average values of methane concentration as well as the maximum and minimum values. The average methane concentration values were calculated for twenty-four hour periods, from 6:00:00 on the current day to 5:59:59 the following day. The maximum and minimum methane concentration values were determined in the same periods.

Figure 4 shows the average, maximum and minimum daily methane concentration throughout the observation period.

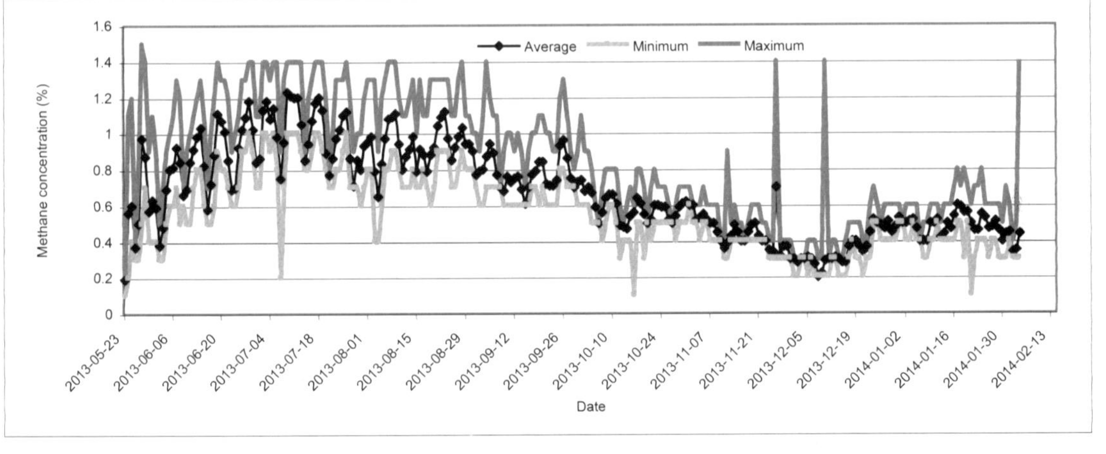

Figure 4. The average, minimum and maximum methane concentration during the observation period.

Figure 4 shows long-term changes in methane concentration and fluctuations associated with the five-day working week (mining was conducted from Monday to Friday). In weekly fluctuations, the highest concentrations of methane occurred from Wednesday to Friday, while the lowest concentrations were generally observed on Sundays.

The average methane concentration at the outlet of the N-6 longwall ventilation area increased steadily from the beginning of the exploitation (from 23.05.2013) to 09.07.2013. The increase in methane concentration was certainly due to the formation of the complete methane desorption zone both in the roof layers and in the floor ones. In the discussed period big differences between the highest and the lowest concentration of methane in weekly periods are visible.

From 09.07.2013, the methane concentration level was constantly decreasing, initially (from 25.08.2013) slowly and then faster. With the increase of exploitation time, differences between the minimum and maximum methane concentrations also decrease. This is a period of exploitation during which an increasing length of the longwall enters under the mined-out part of seam 329/2, 329/1-2.

After 09.12.2013, there was a short period of the average methane concentration increase, followed by its stabilization. This fact is related to a reduction of the air flow rate by about 500 m^3/min (figure 6).

Figure 5 shows graphs of methane concentration and output on respective days of the observation period.

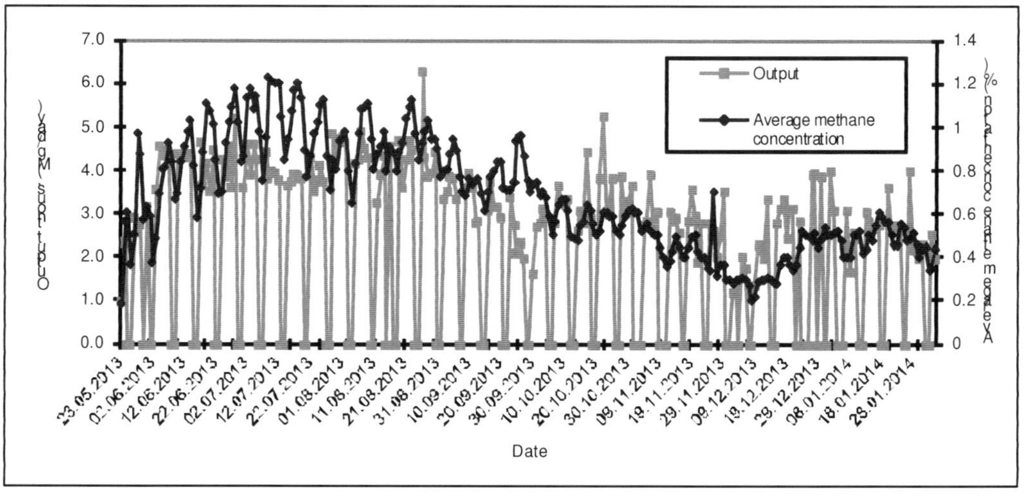

Figure 5. Changes in average methane concentration and daily output during the observation period.

Graphs suggest that the dependence between presented values is not high. This is confirmed by the value of linear determination coefficient between the average methane concentration and output, which amounts to R^2 = 0.179, between daily output and minimum concentration R^2 = 0.115 and between daily output and maximum methane concentration R^2 = 0.169. This means that the variability in the average methane concentration can be explained by the variability of output only in 17.9%, the variability of the minimum methane concentration in about 11.5%, and the variability of the maximum methane concentration in about 16.9%.

Figure 6 shows graphs of air flow rate and the average methane concentration.

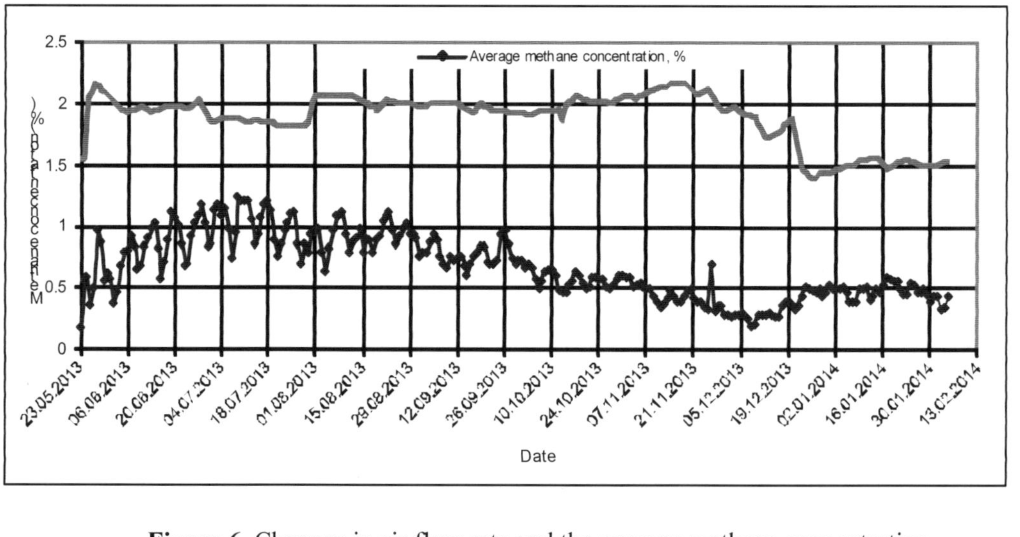

Figure 6. Changes in air flow rate and the average methane concentration.

The value of determination coefficient between air flow rate and the average methane concentration is $R^2 = 0.072$, between air flow rate and the minimum methane concentration $R^2 = 0.065$ and between air flow rate and the maximum methane concentration $R^2 = 0.068$. Values of determination coefficients indicate a very weak dependence of methane concentration from air flow rate. The reason for such low dependence is relatively small variability of air flow rate and at the same time quite significant variability of methane emission to the longwall area.

Based on the above analyses it may be concluded that fluctuations of methane concentration were the effect of several factors (measurable and non-measurable) at the same time, and roles of particular factors in shaping the methane concentration in the longwall area are difficult to quantify unequivocally.

In paper [1] it was found that the average methane concentration on the current day can be described by a linear function of methane concentration on the previous day. It was also ascertained that there exist linear dependencies of the maximum methane concentration from the average methane concentration on the previous day and of the maximum methane concentration from the maximum concentration on the previous day.

Based on statistical analysis of ten longwalls data set in Jastrzebska Coal Company, covering a total of 2239 days, the parameters of linear predictive models were determined for each day of the week (i.e. Monday, Tuesday, etc.) for the average methane concentration in dependence of the average methane concentration on the previous day, the maximum methane concentration in dependence of the average methane concentration on the previous day as well as the maximum methane concentration in dependence of the maximum methane concentration on the previous day. Using the above forecasting models, on the basis of measurement data collected at the outlet of the N-6 longwall in seam 330/2 in the ‘B’ mine, the one-day *ex post* forecasts were carried out and then errors of these forecasts were examined.

4. Forecast of the average methane concentration

The average methane concentration forecast was made for all days of the observation period except for the first one, for which there is no descriptive variable (independent variable), i.e. methane concentration on the previous day. Graphs of the average methane concentrations for measured and forecasted values are shown in figure 7. The comparison of both graphs shows that forecasts exhibit a significant consistency with measurement data.

Figure 8 shows the linear correlation between the average forecasted concentration and the average measured concentration of methane.

In the case of ideal compliance of forecast with measurements, the values of the gradient of a straight line and the determination coefficient are 1.

The equation of the straight line shown in figure 8 leads to a conclusion that the gradient of a straight line adopts the approximate value of $a = 0.97$ and therefore is close to one. On the other hand, the value of determination coefficient is $R^2 = 0.87$, indicating a very strong coherence between forecasted methane concentration and the measured one [12].

For each day of forecast, absolute error was calculated with the accuracy of 0.01% CH_4, i.e. with the accuracy of the average methane concentration value calculation. A summary of the average methane concentration forecast absolute errors is shown in figure 9.

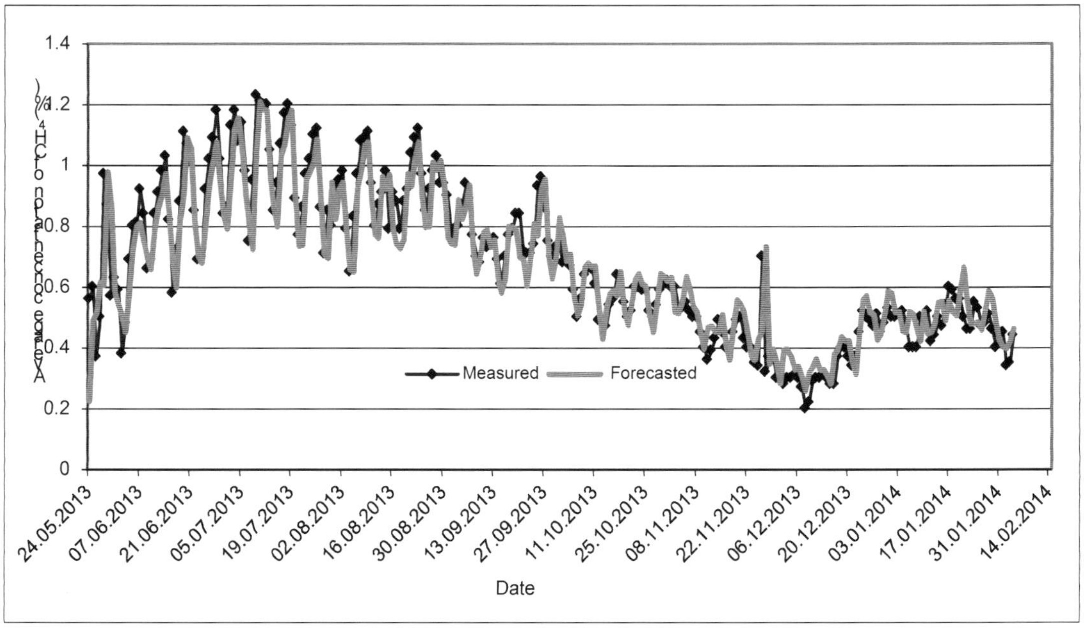

Figure 7. Graphs of the average methane concentration, measured and forecasted.

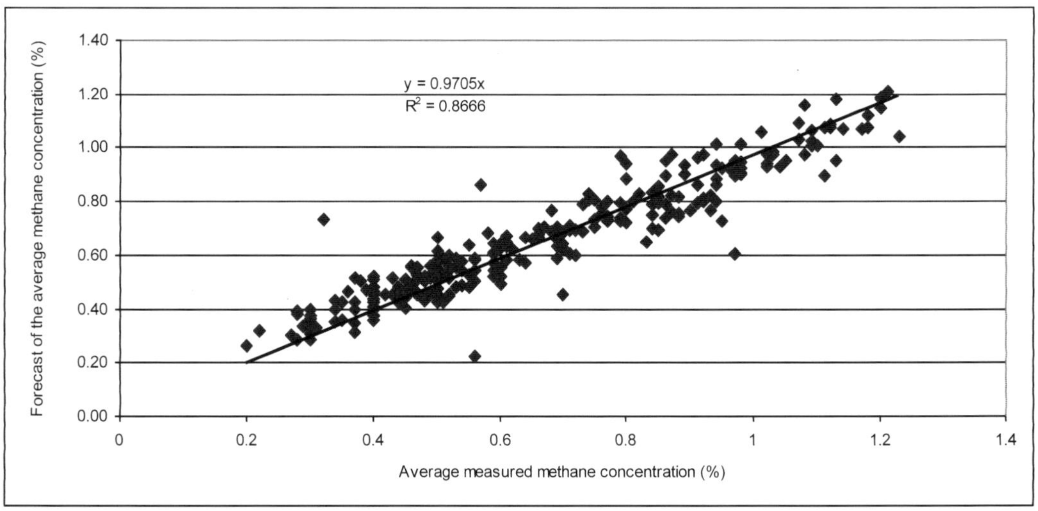

Figure 8. Linear correlation between the average measured and forecasted methane concentration.

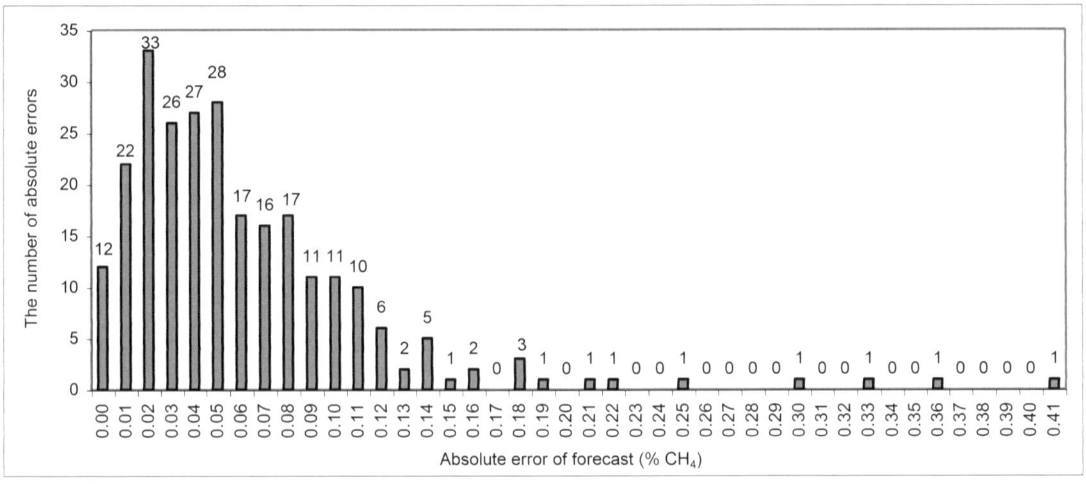

Figure 9. Histogram of quantity of the average concentration of methane forecasts absolute errors.

The horizontal axis specifies the upper limits of absolute errors values intervals, computed with an accuracy of 0.01% of CH_4. For example, the value of 0.1 refers to the left-half open interval of absolute errors (0.09; 0.1). The number of absolute errors is indicated on the vertical axis, and the number of absolute errors in the given interval is shown over individual columns in the graph.

The figure shows that errors above the value of 0.15% CH_4 are accidental errors caused by excessive disturbances in methane inflow into the longwall area or ventilation problems on the day for which forecast was made or on the previous day. The above statement shows an example of the average methane concentrations on 26.11.2013 and 27.11.2013 (figure 7, table 1).

Table 1. Exemplary results of measurements, forecasts and forecasts errors of the average methane concentration.

Date	Methane concentration measured (%CH_4)	Methane concentration forecasted (%CH_4)	Difference (%CH_4)	Absolute error (%CH_4)
24.11.2013	0.35	0.36	-0.01	0.01
25.11.2013	0.34	0.43	-0.09	0.09
26.11.2013	0.70	0.45	0.25	0.25
27.11.2013	0.32	0.73	-0.41	0.41
28.11.2013	0.37	0.35	0.02	0.02
29.11.2013	0.37	0.04	-0.03	0.03

On November 24th and 25th, 2013, the average measured methane concentration amounted to 0.35% CH_4 and 0.34% CH_4 respectively, and forecasted methane concentration amounted to 0.36% CH_4 and 0.43% CH_4. Absolute errors amounted to 0.01% CH_4 and 0.09% CH_4. On 26.11.2013, the average methane concentration suddenly increased to 0.70% CH_4. Since forecast for that day uses the measured concentration of the previous day, forecasted value amounts to 0.45% CH_4. Therefore, absolute error of forecast is large and amounts to 0.25% CH_4. On the next day (27th November) the measured methane concentration was reduced to 0.32% CH_4. Forecast for this day amounts to 0.73% CH_4 and is significantly overstated because it bases on the value of methane concentration from the previous day (0.70% CH_4), much higher than on the few days before. Hence the absolute error of forecast for this day amounts to 0.41% CH_4. On the next two days the disturbances in methane concentration are relatively small, which makes absolute errors of forecast also small.

Figure 10 shows the total percentage of absolute errors from 0.00% to the value specified on the horizontal axis.

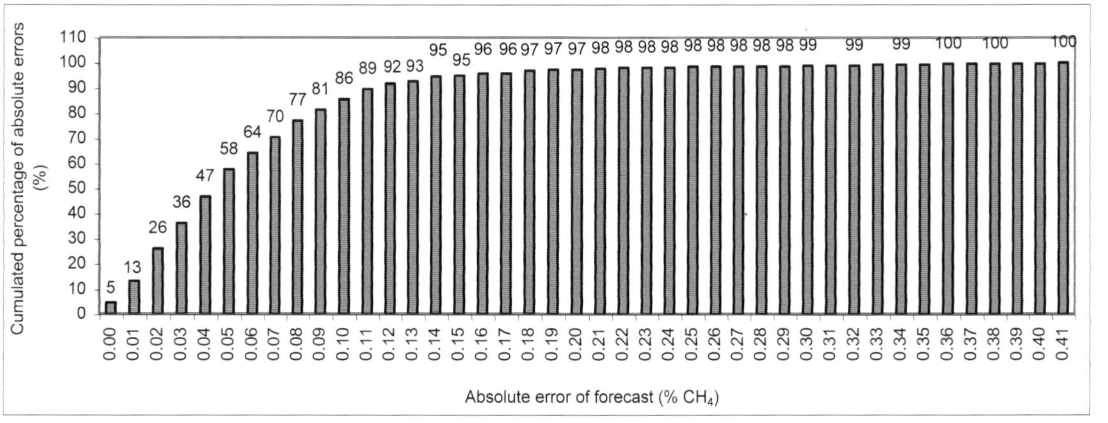

Figure 10. Histogram of the cumulated percentage of absolute errors of the average methane concentration forecast.

The graph in figure 10 shows that in the interval of 0.00% CH_4 to 0.10% CH_4 there are 18% of absolute errors of forecast, in the interval of 0.00 CH_4 to 0.10% CH_4, there are 86% of absolute errors of forecast, and in the interval of 0 to 0.15% CH_4 there are as many as 95% of absolute errors of forecast. The remaining 5% of forecast errors is due to accidental, large disturbances in methane concentration.

Analyzes performed in paper [12] have shown that residuals of prognostic models do not have a normal distribution. Accordingly, the theoretically calculated forecasts errors that are correct in the event of a normal distribution of residuals of the model would be burdened. Therefore, the expected accuracy of the proposed prognostic method will be possible to assess only after performing numerous forecasts and evaluating their errors.

For engineers of mine ventilation divisions who make use of forecast, the important issue is a relative error of forecast with a predetermined value, which they consider acceptable. In the following part of this paper it was assumed that this error amounts to 20% of forecasted average value of methane concentration.

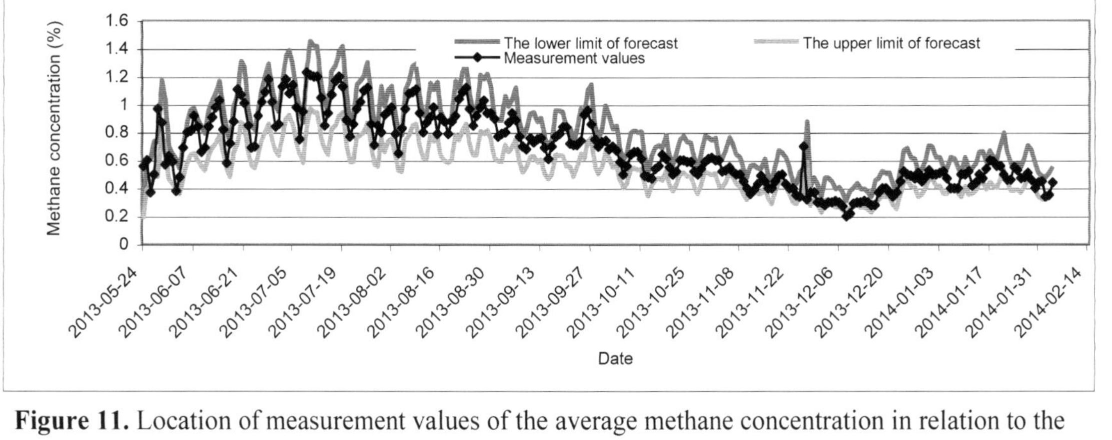

Figure 11. Location of measurement values of the average methane concentration in relation to the 20% deviation interval from forecasted average methane concentration.

In figure 11, the green line (the lower limit of forecast) represents 80% of forecasted average methane concentration and the red line - 120% of this value (the upper limit of forecast). The black line connects the measurement values of the average methane concentration.

Calculations have shown that in forecast interval there are 232 of the total 256 values of the average measured concentration, which is approximately 90.3% of all values.

For safety reasons it is important to know the number of exceedances of the upper forecast limit by the average methane concentration. The calculations show that such relationship occurred in only 10 cases, which accounts for 3.9% of cases.

The above analysis makes it clear that the accuracy of the average methane concentration forecasts is very good.

5. Forecast of the maximum methane concentration

Based on prognostic equations in paper [1], two types of the maximum methane concentration forecasts were performed. The independent variable in forecast 1 is the average methane concentration on the previous day, and in forecast 2 it is the maximum methane concentration on the previous day.

Graphical comparison of the maximum methane concentration measurement values with values of forecast 1 is shown in figure 12, and with values of forecast 2 - in figure 13.

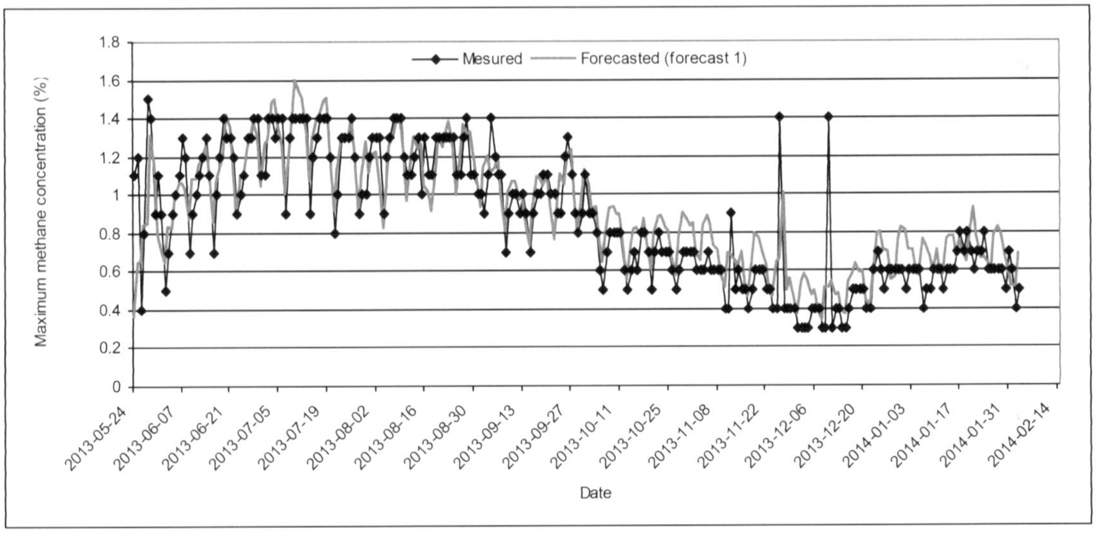

Figure 12. The values of maximum methane concentration measured and forecasted on the basis of the average methane concentration on the previous day (forecast 1).

Comparison of figures 12 and 13 shows that the maximum methane concentration values obtained from forecast 1 are slightly higher than from forecast 2. This observation may be confirmed by the fact that the sum of forecasted values for the whole analyzed period is 231 according to forecast 1, and 216 according to forecast 2, while the sum of the maximum concentration measurement values amounts to 221. The obtained maximum methane concentration values according to forecast 1 are underestimated in 88 cases and overestimated in 168 cases, while according to forecast 2 forecasted values are underestimated in 129 cases and overestimated in 127.

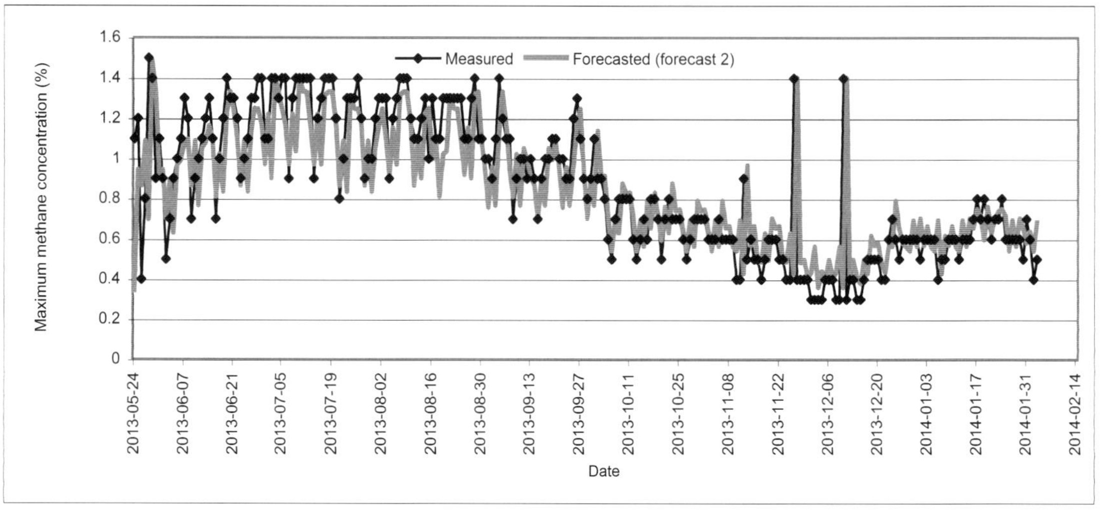

Figure 13. The values of maximum methane concentration measured and forecasted on the basis of the maximum methane concentration on the previous day (forecast 2).

Figure 14 shows the distribution of absolute errors of the maximum methane concentration forecast 1, and Figure 15— the distribution of absolute errors of forecast 2.

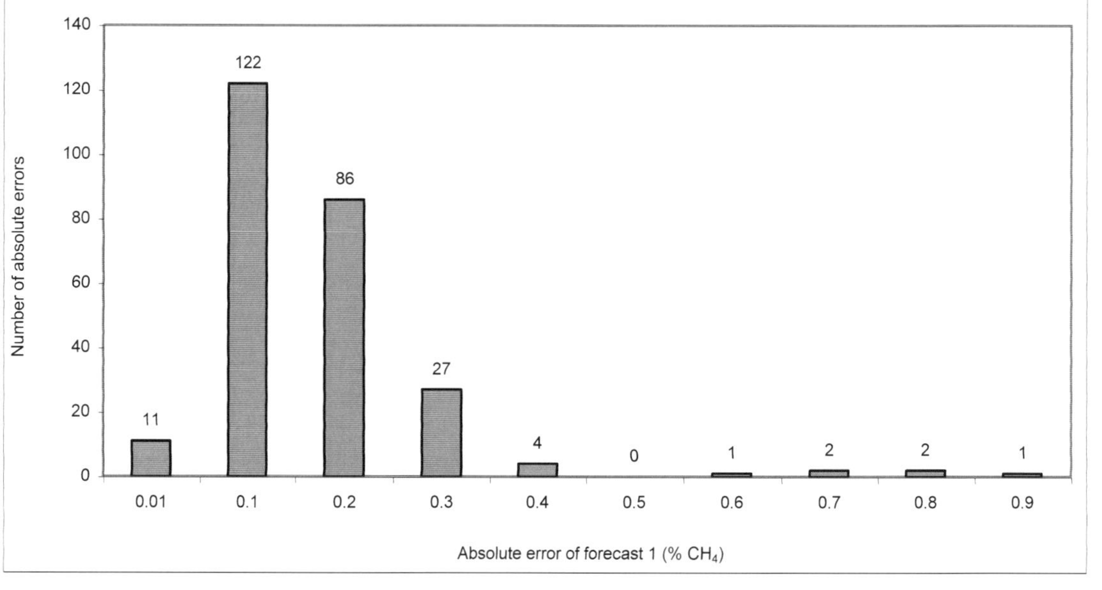

Figure 14. Histogram of absolute errors quantity for the maximum methane concentration forecast 1.

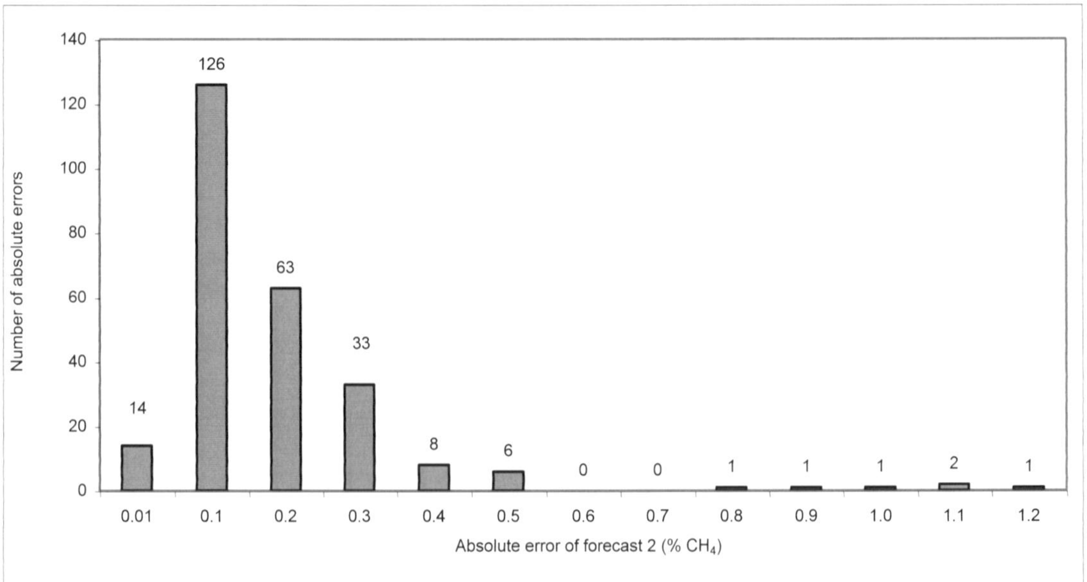

Figure 15. Histogram of absolute errors quantity for the maximum methane concentration forecast 2.

Forecast 1 is characterized by a lower number of absolute errors to the value of 0.1% CH_4 than forecast 2. In forecast 1, absolute errors do not exceed the value of 0.9% CH_4, while in forecast 2 they reach the value of 1.2% CH_4.

Figures 16 and 17 show that absolute errors till the value of 0.1% represent 52% of all errors of forecast 1and 54.7% of all errors of forecast 2. On the other hand, absolute errors from an interval of 0 to 0.3% represent 96.1% of all errors of forecast 1 and 92.2% of errors in case of forecast 2. This means that absolute errors greater than 0.3% CH_4 represent 3.9% of the total number of errors in case of forecast 1 and 7.8% of all errors in case of forecast 2.

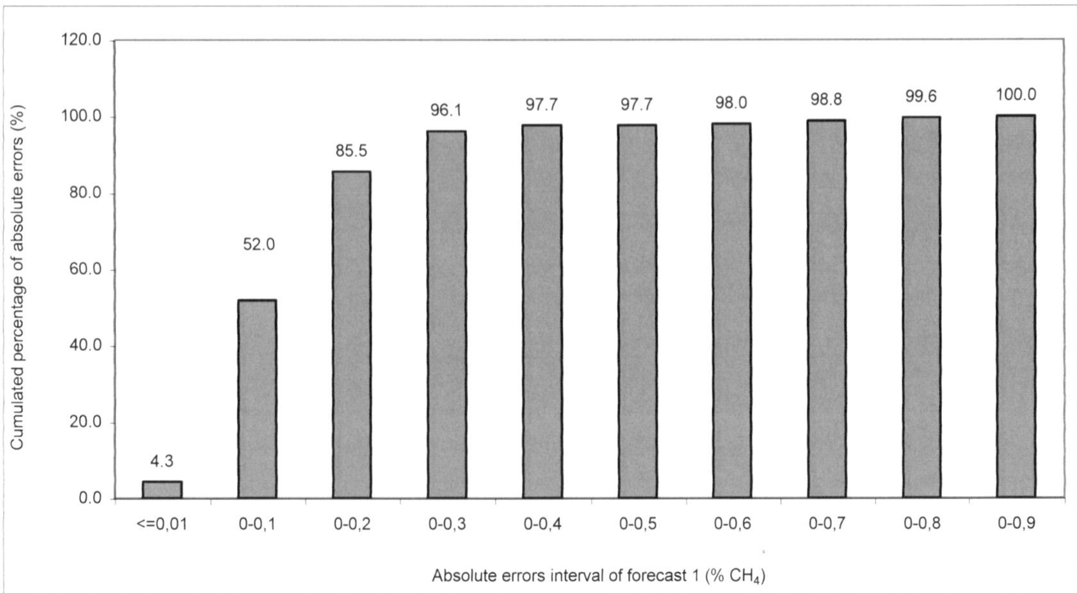

Figure 16. Histogram of the cumulated percentage of absolute errors of the maximum methane concentration forecast 1.

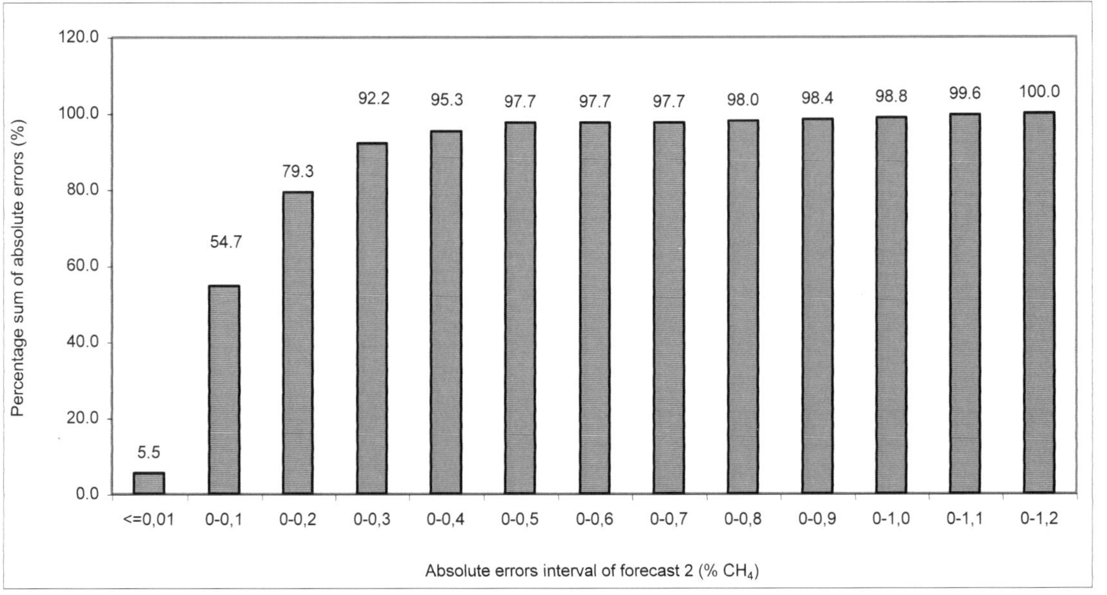

Figure 17. Histogram of the cumulated percentage of absolute errors of the maximum methane concentration forecast 2.

Figures 18 and 19 show forecast interval [$0.8S_{max}$, $1.2S_{max}$], where S_{max} represents forecasted maximum methane concentration. The value of $0.8S_{max}$ is called the lower limit of forecast, and the value of $1.2S_{max}$ is the upper limit of forecast. Figure 18 refers to forecast 1 and figure 19 - to forecast 2.

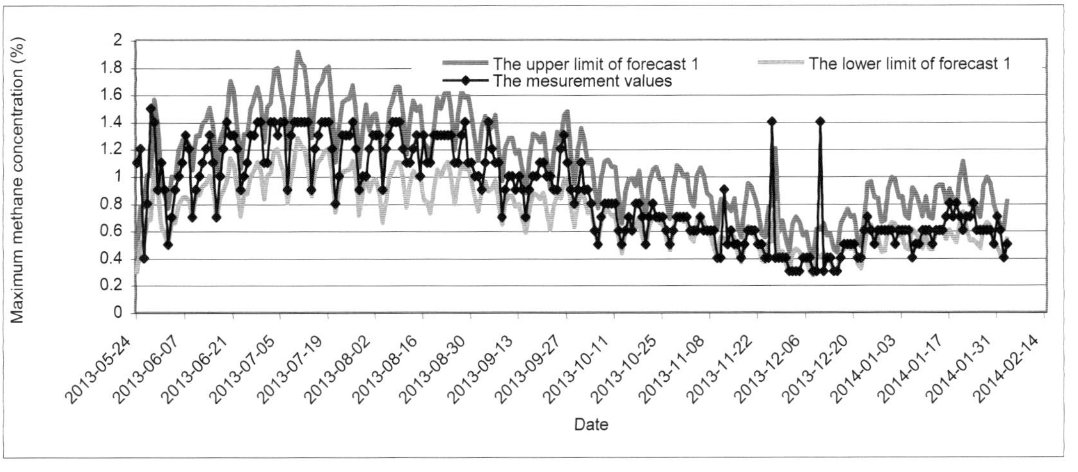

Figure 18. Location of measurement values of the maximum methane concentration in relation to the 20% deviation interval from forecast 1 of maximum methane concentration.

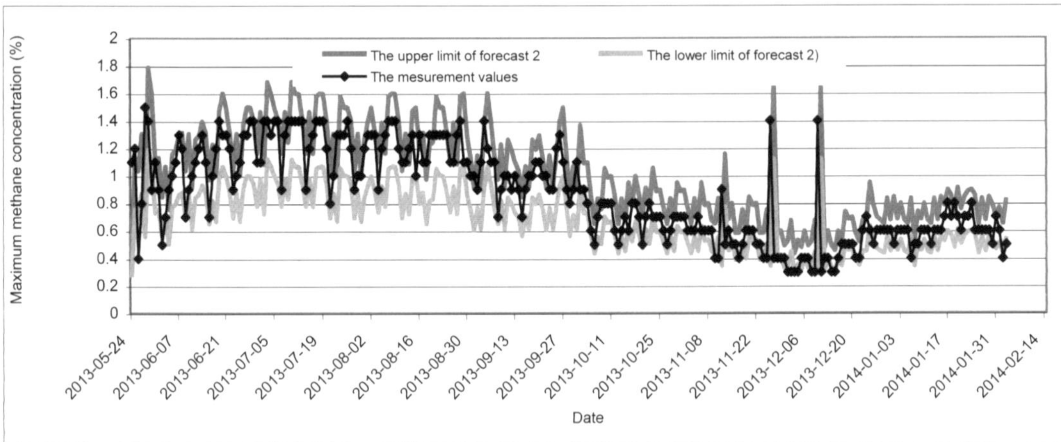

Figure 19. Location of measurement values of the maximum methane concentration in relation to the 20% deviation interval from forecast 2 of maximum methane concentration.

Within the limits of forecast 1 there are 188 cases of the maximum methane concentration measurement values, and within the limits of 2-interval forecast - 182 cases, corresponding to respectively 73.4% and 71.1% of the total number of 256 cases.

For security reasons, underestimation of forecast is more unwelcome than its overestimation. Over the upper limit of the 1-interval forecast occurred 15 cases of the maximum methane concentration measurement values, and over the upper limit of 2-interval forecast occurred 42 cases. This represents 5.9% and 16.4% of the total number of cases, respectively.

The above specification shows that the maximum methane concentration forecast that uses the average methane concentration on the previous day (forecast 1) is in case of the N-6 longwall more accurate than forecast that uses the maximum methane concentration on the previous day.

6. Recapitulation and conclusions

Exploitation of the N-6 longwall in seam 330/2 in the 'B' coal mine owned by Jastrzebska Coal Company was conducted in complex geological conditions. The roof and floor layers of seam had variable thickness as well as mineralogical and petrographic composition. In seam 330/2 there were four layers of gangue.

In the initial phase of exploitation, the longwall was situated under the unexploited part of seam 329/1, 329/1-2. The longwall advance revealed the rising trend of methane concentration at the outlet of the longwall area. That trend was connected with the expansion of methane desorption zones in rocks above and below the exploited seam. After mining out about 220 m of panel length, the longwall was gradually entering under the abandoned workings in seam 329/1, 329/1-2, which changed the trend of methane concentration to descending. Due to exploitation works conducted five days a week, over the whole period appeared fluctuations in methane concentration.

This paper presents the application of methods developed in paper [1] for forecasting the average and maximum methane concentration at the outlet of the N-6 longwall area in seam 330/2.

Methane concentrations were analyzed by continuous measurements performed by methane concentration sensor located at the outlet of the longwall ventilation area, which is an element of the 'B' mine telemetric system. On the basis of these measurements, the average methane values for 24 hours (from 06:00:00 to 06:59:59 the following day) were calculated and the maximum and minimum methane concentrations were determined. Measurement error of methane concentration amounted to 0.1% CH_4. Forecasts covered a 256 day period of the longwall exploitation.

Analysis of forecasts results allowed for the following conclusions:

- The average forecasted concentrations in 86% cases of the total number of forecasts (256) differed from the measured value by no more than 0.01% CH_4 and in 95% cases by no more than 0.15% CH_4.
- In the interval of 80% to 120% of the average methane concentration forecasted value occurred 90.3% of the measured average methane concentration cases. In 3.9% of cases the average measured concentration was higher than 120% of the average forecasted concentration.
- The results of forecast lead to the conclusion that forecasts for the average values of methane concentration show a very good accordance with the average methane concentration measured values.
- The maximum forecasted methane concentrations on the basis of the measured average values of methane concentration on the previous day (forecast 1) differ in 52% of cases from the measured values by not more than 0.1% CH_4, which corresponds with the measurement error. In 85.5% of cases the forecast error is no greater than 0.2% CH_4, and in 96.1% of cases this error amounts to no more than 0.3% CH_4.
- In the interval of 80% to 120% of the maximum methane concentration values, calculated according to forecast 1, occurred 73.4% of the measured maximum methane concentration cases. In 5.9% of cases the maximum measured concentration was higher than 120% of forecasted values.
- The maximum forecasted methane concentrations on the basis of the measured maximum values of methane concentration on the previous day (forecast 2) differ from the measured values by not more than 0.1% CH_4 in 54% of cases. In 79.3% of cases absolute error of forecast is no greater than 0.2% CH_4, and in 92.3% of cases this error amounts to no more than 0.3% CH_4.
- In the interval of 80% to 120% of the maximum methane concentration value calculated according to forecast 2 occurred 71.1% of the measured methane concentration cases. In 16.4% of cases the measured maximum methane concentration took higher values than 120% of the maximum methane concentration calculated according to forecast 2.
- Both the results of forecast 1 and forecast 2 of the maximum methane concentration can be assessed as good. Comparison of the results of forecast 1 and forecast 2 showed that forecast 1 was slightly more accurate.

References

[1] Badura H 2013 Metody prognoz krótkoterminowych stężenia metanu na wylotach z rejonów ścian zawałowych w kopalniach węgla kamiennego (Gliwice: Wydawnictwo Politechniki Śląskiej)

[2] Rozporządzenie Ministra Gospodarki z dnia 28 czerwca 2002 r. (Dz.U. nr. 139, poz. 1169 wraz z późniejszymi zmianami) w sprawie bezpieczeństwa i higieny pracy, prowadzenia ruchu oraz specjalistycznego zabezpieczenia przeciwpożarowego w podziemnych zakładach górniczych

[3] Boxho J, Stassen P, Muecke G, Noack K, Jeger Ch, Lescher L, Browning E J, Dunmore R and Morris I H 1980 Grubengasabsaugung (Essen: Verlag Glueckauf GmbH)

[4] Karacan C O, Ruiz F A, Cote M and Phips S 2011 Coal mine methane: A review of capture and utilization practices with benefits to mining safety and to greenhouse gas reduction International Journal of Coal Geology 86 pp 121-56

[5] Kozłowski B and Grębski Z 1982 Odmetanowanie górotworu w kopalniach (Katowice: Wydawnictwo „Śląsk")

[6] Krause E and Łukowicz K 2000 Dynamiczna prognoza metanowości bezwzględnej ścian. Poradnik techniczny (Katowice-Mikołów: Wydawnictwo Głównego Instytutu Górnictwa)

[7] Fluegge G 1971 Anwendung der Trogtheorie auf den Raum der Zusatzausgasung Glueckauf-

Forschungshefte 32 pp 122-9
[8] Koppe U 1975 Der ausgaungsgrad von Beglleitfloezen im Liegenden der flachen Lagerung Glueckauf-Forschungshefte 4 pp 136-44
[9] Noack K 1970 Untersuchungen ueber Form und Groesse des Ausgasungsraumes um Abbaubetriebe in flacher oder maessig geneigter Lagerung des Ruhrkarbons Glueckauf-Forschungshefte 3 pp 121-32
[10] Noack K 1976 Der Einflus von Streblaenge, Abbaugeschwidigkiet und Versatzart auf die Ausgasung Glueckauf 20 pp 1156-9
[11] 2013 Dokumentacja techniczna ściany N-6 w pokładzie 330/2 w KWK 'B'
[12] Zeliaś A 2000 Metody statystyczne (Warszawa: Polskie Wydawnictwo Ekonomiczne)

Institute of Physics
Dirac House, Temple Back
Bristol BS1 6BE UK

ISSN: 1757-8981
ISBN 978-1-5108-5267-9

CURRAN ASSOCIATES INC.
proceedings.com

International Conference on Materials Engineering and Applications (ICMEA 2018)

IOP Conference Series: Materials Science and Engineering
Volume 348

Bali, Indonesia
14 – 16 January 2018

IOP
Institute of Physics